HELP YOUR CHILD WITH SCIENCE

HELP YOUR CHILD WITH SCIENCE

Steve Pollock
and
Julian Marshall

BBC BOOKS

For Laura, Jamie and Lesley who helped me. S.P.

With thanks to Rachel, Anna and Ruth. J.M.

Published by BBC Books, a division of BBC Enterprises Ltd
Woodlands, 80 Wood Lane, London W12 0TT

First Published 1992

ISBN 0 563 36215 4

Illustrations by Richard Geiger

Set in Century Schoolbook by Goodfellow & Egan Ltd, Cambridge
Printed and bound in Great Britain by Clays Ltd, St Ives plc

Contents

Acknowledgements

We would like to thank our colleagues George Auckland and Jenny Stevens for supporting the project from its initiation, Suzanne Webber for setting us on the right road and Sally Crawford for her editing of the manuscript. Also all those whom we contacted for their views and information particularly Sue Pringle, University of Bristol, the SPACE project team, Dr Joan Solomon, SHIPP project, and Anne Watkinson, coordinator of the ASE Primary Science – a shared experience project for parents.

About the authors

STEVE POLLOCK has written widely for adults, children and teachers and has had considerable experience in developing learning materials and activities both for schools and in informal situations out of schools. Before joining the BBC as an Education Officer he was responsible for visitor education at the Natural History Museum in London. He has broadcast regularly on aspects of science and the environment and was elected a Fellow of the Institute of Biology for his contribution to biological education.

JULIAN MARSHALL has over the past four years concentrated on developing training materials for teachers to help them develop science teaching in both primary and secondary schools as part of the new national curriculum. In order to do this he has drawn on experience as a research engineer in telecommunications and as a science teacher.

Authors' note

This book is for anybody who is responsible for bringing up children, particularly children between the ages of three and twelve. It has been written in the hope that it will help you take and make opportunities to inform, encourage and share in, with your child, an exploration of the world around you: an activity known incidentally as science. As the book will show you, you need know no formal science yourself in order to help your child and to become a partner in his or her discoveries.

The book puts such activities into the context of science as a core part of the new national curriculum, sharing the honours with English and maths. It fills in the background to the curriculum changes as they affect primary schools since, for the majority of primary schools, this emphasis on science will be new.

During the course of compiling *Help your child with science* the authors have canvassed the views of children themselves as voiced during numerous visits made to schools (in their capacity as BBC education officers) as well as by their own children.

Help your child with science is a companion volume to the BBC TV series and to the BBC publications *Help your child with maths* and *Help your child with English*. It is, in its entirety, addressed to both boys and girls.

INTRODUCTION

What's going on? The word 'science' is being spoken of in the same breath as the word 'education'. What *is* going on? It's quite simple. All children are now taught science at school from the age of five.

This in many ways momentous event came about on 1 August 1989 as a result of the Education Reform Act. All children between the ages of five and six attending state schools are now taught science as the first step in the implementation of the new national curriculum. At the end of the summer term of 1991 those same children will have been assessed. From now on the intention is that every child will follow the same carefully laid out educational path, using a framework which is common to all schools.

In Northern Ireland, similar arrangements have been made and started in the autumn term of 1990 under the Northern Ireland curriculum council.

Similarly, in Scotland, science remains an essential part of the curriculum in primary schools and forms part of environmental studies within the new 5–14 development programme.

Independent schools, although not bound by the national curriculum, have been active in the consultation processes that led to its being set up and eager to incorporate it into their own programme of science education.

So as a result of all this, science has now become a central part of every child's education, carrying equal status with reading, writing and maths. From the age of five every child, including children with special needs, will be involved in activities which lay the foundations for ideas about energy, electricity and evolution. From an early age children will be developing skills that are important to scientific exploration such as measuring, testing and putting forward and communicating ideas about the way the world around us works.

Wonderful for the children. How about the parents?

Note: For convenience, when we have made reference to a particular set of ideas in science education, we have based our remarks on the National Curriculum for England and Wales.

Why science?

There are many answers to this question. But first of all perhaps we should ask 'why the question?'

Well, for a start, government legislation now decrees that by sixteen *all children* are entitled to take their GCSE (Standard Grade in the case of Scotland) in science. In most cases, this means that science will take up a fifth of children's time in school. What role does science play in our lives that it should warrant so much attention?

Here is a list of some of the arguments 'for' science. We hope that by the time you have read this book, with your child, that you will want to add some more of your own.

- Science is now a central part of the school curriculum for children of all ages.
- Science directly affects our quality of life – where would we be without drinkable tap water or penicillin and anaesthetics when we need them?
- Decisions we make at a personal level need thinking through using 'scientific ideas' – is double glazing worthwhile? Can we afford it? Will it help us save on central heating if we have to pay a carbon tax?
- Being scientifically informed ensures that we can be properly involved in the democratic process. Science discovered nuclear fission and how to extract fossil fuels. What principles do you use when considering the merits of using their energy as our power? If we ban both do we realise what kind of world we would be returning to?
- Science is an extension of our natural curiosity to explore and make sense of our surroundings. Einstein called it 'a refinement of common sense'.
- Science no more owns our world than doctors own our bodies: the ideas we ourselves have about the world we live in are intrinsically as interesting and relevant as the ideas we have about painting, music, the latest TV programme or our favourite hobby.
- A final key point: no one, least of all the people who make up the national curriculum council, expect children to develop their ideas in science solely at school, in isolation from the rest of their experience – so children's experience at home really does count.

We hope this book will go some way to helping you increase your own level of understanding and enjoyment of science – and – since some of it is actually quite interesting and none of it involves test tubes – that you will want to share this new understanding with your child and make him or her a partner in your new discoveries.

What is science?

If you ask a scientist this question you will probably get a shifty look. Opinions vary and they change with time. Indeed, scientists still argue about the precise meaning; some may not want to comment at all in case they get it 'wrong'. The most lucid explanation ever heard by one of the authors (JM) came in fact from a thirteen-year-old girl during a schools visit: 'it's about (*pause*) finding out (*pause*) about everything'. That statement contains two important ideas (whether intended or not). Firstly, science finds out about the world around us and secondly, it's about how we carry out that finding out.

The national curriculum reflects this in stating, 'pupils should use a variety of techniques to investigate their hypotheses'. Thinking up a hypothesis means using your imagination. The imagination thinks up a way to describe what 'might be' and, if so, 'what if'. These imaginative skills are not at all unique to science.

ONE MAN'S IMAGINATION

During the nineteenth century an Austrian scientist named Ludwig Boltzman decided that it would be a good idea to try and explain the way gases behave by imagining that everything was made up of tiny particles. In Boltzman's imagination, these particles were too small to see and in a gas they flew about continuously bumping into each other quite randomly. We can (and do) use this simple idea about imaginary particles to explain what we actually see and feel – but someone had to think it up first.

This particular hypothesis has many everyday uses. The tyres on a bicycle feel hard because you have pumped lots of these particles into the tube. They fly around banging on the inside walls of the tyre, knocking it outwards. If you let some air out then there are fewer particles inside the tyre so the inside walls of the tyre get hit less often and the tyre feels squashy.

If you then extend this theory using some elaborate mathematics to help (as Boltzman did) you end up with the theory on which all the major developments in industrial chemistry this century have been based. This is the kinetic theory of matter or if you're at university, what is known as statistical mechanics. It's sad to think that at the time everybody thought Boltzman was completely mad.

PUTTING THEORIES TO THE TEST

Science pays careful attention to the methods which it uses in finding things out, to the extent that scientists themselves are the first to acknowledge that an accepted theory could be wrong. Michael Faraday a blacksmith's son, and one of the greatest experimental scientists who ever lived, said at the end of one of his Royal Institution lectures, 'the man who is certain he is right is almost sure to be wrong; and he has the additional misfortune of inevitably remaining so . . . ever since the world began opinion has changed with the progress of things'. Faraday was well qualified to make such a bold statement as it was he who was able to make sense of the strange effects which later led to the generation of electricity and the invention of the electric motor.

He was right too. At one time it was fashionable to believe that the earth was at the centre of the universe and that everything including the sun circulated round it, an idea founded by the ancient Greek philosophers 1700 years earlier. In the seventeenth century, Galileo Galilei turned his new telescope to look at the stars and made some observations which supported a completely different set of ideas: that is that the earth goes round the sun. So threatening did the church and State find this suggestion – it did after all mean that the rules on which everything was based were wrong – that Galileo was put on trial and made to withdraw his comments. Later, Isaac Newton used this same evidence and more in developing his laws of motion, a set of laws which scientists will choose to use in preference to all the earlier ones in planning the next space shuttle launch.

How you can help your child

The desire to find out is within all of us. Much of the way children go about finding out resembles experiment and in this respect young children are already practising scientists. Primary schools seek to build on and extend this natural curiosity. We can help by providing children with a range of safe activities which allow them to continue their explorations and by being an audience for them to explain their ideas to.

It is of prime importance to listen to your child. Knowing that someone is truly listening means we can relax and explain what we mean more easily. Children's ideas about the world change – a similar process to what happens in the scientific mind. Many of the ideas and explanations your child gives may seem wrong to you. They are, however, probably consistent with the child's own experience, and therefore probably good science. Let them make their own discoveries: it is of course an advantage if you are on hand to offer any guidance you can (this might save them duplicating thousands of years of human exploration) and (most of all) to provide them with opportunities to help them develop their ideas over time.

Listening well does not always mean setting aside special time to be with your child while you both think about 'science' together; on the contrary, as you will see from this book, it means listening to your child both at home and out of it in ordinary, everyday situations: while they are having a bath; while they are getting ready for school; while you are both waiting for the bus or wheeling a trolley around the supermarket.

The other important thing is to talk to your child. Conversation of any kind gives children the opportunity to verbalise their ideas and thoughts. They can then try them out and learn to refine them. The home environment is actually a more relaxed place for children to ask questions and develop conversation than, say, nursery school. This is no surprise when you think about how many children have to share their time with a teacher.

Through conversation, you supply your child with plenty of new words to extend their understanding. This is particularly important while they are undertaking any activity. For example, in simple terms, if they move a toy car under a bridge you need to give them the vocabulary to help them understand the concept of under and then over. Without the words a child has no label and therefore no means to communicate an idea. These words are very useful later in making connections and modifying ideas. Throughout learning (indeed throughout life) vocabulary and experiences go hand in hand and for science this is particularly important. Never be afraid of using words which seem too complex: most young children can get their minds and tongues around tricky names; they are perfectly at home, for example, knowing the names of the different dinosaurs.

Encourage your child to ask questions. This is an important way for children to learn about the world. Accept that you cannot know everything but look for ways to help. For example:

- direct questions, e.g. 'what is a cloud?' can be looked up in a book;
- some questions have no answer but they can still be responded to, e.g. 'if God made the world who made God?';
- some questions are difficult to answer because your child simply does not have the experience or conceptual understanding to make sense of your answer to a question such as 'how does a computer work?'

Some of the above questions can be broken down into a series of sub-questions that can be explored through experiment, for example, 'why do ships float?' You could turn this into a series of activities which would help your child gain actual experience of concepts. Carrying out activities such as those on pages 22 and 23 would help answer this particular question.

INTRODUCTION TO THE ACTIVITIES

In this part of the book you will find activities that you can carry out with your child around the house, around the garden or in the immediate environment. All of them tap into a child's natural curiosity, help him or her to overcome the frustration of 'not knowing' and help you as a parent to help your child.

Activities include those centred around the following places:

Your bathroom	Your nearest town
Your kitchen	Your nearest bit of countryside
Your garden (or park)	At the seaside

In addition, following on from these location-based activity sections, you will find three topic-based ones on themes that are popular in schools. These should provide you and your child with further starting points for exploring and enjoying science.

Topics include:

Our pets
Ourselves
Toys and games

To help you further, the following 'information' headings are common to each of the sections:

Have you ever noticed? – this covers all the things you may have noticed in your everyday environment but might want to know more about – why milk boils over so fast, for example, or why a bicycle tyre gets harder to pump the longer you go on.

Starting points – these include many of the things you may have done many times before with your children – gone to the shops or visited the park. How do you start to make it new and exciting; what do your children already know?

What's going on? – this is our special 'explanatory' section for parents who welcome the idea of going further – to satisfy their own curiosity as well as that of their child. It consists of explanations of the phenomena that you, as a parent who is having to answer all those questions, may appreciate.

Something to do – how to help your children make the most of the activities. Once you have got the children started their natural enthusiasm will carry them along: here are ways to intervene and ways to adapt the activities so that children get the very most out of them.

Things to explore – How to 'find out' with a list of helpful suggestions.

Making connections – everything that children learn in this book can be applied to lots of other situations they will find themselves in: building a soap bubble machine, for example (page 105), will lay the ground for teaching the principles of molecular attraction when they come to study physics and chemistry.

National curriculum notes – throughout each section you'll find short notes that remind you that the things you are encouraging your children to do and to find out are precisely those that the national curriculum guidelines indicate.

Safety first – a reminder of what to do and what not to do whenever we mention a potential hazard like water, traffic on the roads, or electricity.

We hope you and your child will do many of the activities just for the fun of it – a very good way to learn. But remember that every time you do carry out any of these activities you – and your child – will be taking part, either partly or wholly, in what is known as the 'scientific process':

Observing – looking attentively at the world around you in a way that leads to finding out and making new connections.

Asking questions – what is it that causes something to happen?

Making a hypothesis – providing an explanation for why it happens that way.

Experimenting – testing if your prediction about the way it happens is correct.

Interpreting – using the results of the experiment to decide whether or not your hypothesis was correct.

Communicating – telling somebody else what happened and finding a way of showing them.

Whatever you decide to do together, make sure the mood is right for you and your child. Always wait for interest and enthusiasm then build on it. The activities should always be fun and never forced. Before you start an activity:

- judge the mood of the moment – this is crucial to success;
- try to let your child take the initiative;
- make the most of the occasion and of your child's observations;
- enjoy yourselves – what's the point otherwise!

Science in the Bathroom

Parents seem to spend quite a lot of their time telling children to 'wash'. Here are ways to make this most basic activity – necessary for health and hygiene – a time for exploration and discovery too.

Maybe children would be more civilised if they had some insight into the reasons behind some of our bathroom routines. They might take some of the responsibility for making it a safe and pleasant place (there's never any harm in hoping). But beware this two-edged sword: how prepared are we for questions that challenge our own habits?

It's in the bathroom – the place where we indulge in our most personal of habits – that children always seem to want to ask those particularly awkward questions about ourselves and our bodies. Sometimes, our experience of being adults in an adult world can make us over-sensitive to these innocent enquiries. Here is a story. One parent was asked by her child 'where her best friend came from' and felt obliged to launch into a carefully prepared speech on sex and babies. Her child, taken aback by the extent of her mother's response, made it clear she merely wanted to know where Sheffield was! You can always get more information about the nature of your child's question by asking a question in return – just to check what your child really wants to know. Above all, don't duck. Dealing with these questions as soon as they are asked builds trust and an honest relationship with your child. With luck it might mean children are less likely to ask such questions at even more embarrassing moments in the future – but of course there's no guarantee of that.

Starting points

You can do a lot in a bathroom besides wash. Here we introduce the subject of skin care (scientifically, of course), evaporation, condensation, the principles of floating and sinking, safety and why the bathroom light is 'different'. It's also worth noting that in many primary schools 'Ourselves' is often used as a study theme, particularly in years one and two. You'll find more on this subject in the *Ourselves* section on page 92).

Have you ever noticed?

- ▶ How the skin on your fingers gets shrivelled like prunes.
- ▶ How cold you feel when you get out of the bath.
- ▶ That the light switch in the bathroom is different from those found in other rooms.
- ▶ That life is much more comfortable when you have a bath mat.
- ▶ The way mirrors and wall tiles mist up.
- ▶ That sometimes a sponge might float and sometimes sink.

WHY HAVE MY FINGERS GONE ALL WRINKLY?

Maybe it was just one of those things but if you have time it's worth talking about – after all your child has chosen to share their observation with you. Questions you might think about together include: has it happened before? Does it happen only to fingers? What about your own fingers, does it ever happen to them too? More relevant perhaps 'what do they feel like?'

Over a period of time you can gradually build up the idea together that the wrinkly finger phenomenon seems to happen in water, especially hot water. Needless to say we would never go out of our way to prove the point, young skin being what it is and hot water being potentially dangerous, but basically the effect we are talking about is skin damage. What's happening is that the hot soapy water is washing off not just dirt but also surface lumps of dead skin along with your skin's own protective layer of natural oils which come from the oil glands beneath. This allows the water to attack the underneath layer of skin, which swells and wrinkles along the line of the oil glands. You could also talk about whether or not you or your skin likes it and agree that perhaps it's not such a good thing.

Making connections

After your wrinkly finger investigation you might try and make connections with the use of hand creams (which add a layer of oil) and how to avoid getting 'prune fingers'. Talking about the way water seems to damage things in general also helps make connections. You could observe the effect of water on decorative surfaces and the need to use different types of decorations in the bathroom to those in other rooms in the house. What about the use of sealants around the bath and that nasty stain where the tap's been dripping? More connections include why you need to paint window frames, not leaving toys outside, what happens to wet books, wet shoes, the poor old car and the whole issue of keeping water in its place.

What's going on?

This helps children develop the idea that things are constantly changing. Needless to say, this can lay the foundations for important ideas about the behaviour of materials later – not to mention keeping children from splashing water about quite so freely.

► Sometimes they change back again (your fingers smooth out); sometimes they don't (that damp patch under the sink will need new floor covering). ◄

QUICK, GET DRY AND PUT YOUR PYJAMAS ON BEFORE YOU GET COLD!

Sometimes I don't think my five-year-old believes me but I do remember how reluctant I used to be about getting out of the bath. I was also confused by my mother's insistence that getting dry would relieve my problems, after all wasn't it the hot water that was keeping me warm? When the same problem occurred on the beach I was more confident in getting dry as a strategy because sea water was clearly cold but I was still bemused at the advice that keeping your shoulders underwater would be warmer than not – it did however work.

What's going on?

The effect is due mainly to two things. Firstly, when water evaporates from your shoulders it turns from liquid droplets into a gas. This becomes part of the air around you but in order to change state it needs the warmth (energy) of your skin to do so. So when the water leaves you it takes your energy with it, leaving you cold. (This is a phenomenon known as the latent heat of vaporisation.)

Secondly, the hot bath has encouraged lots of tiny blood vessels to open up near the surface of your skin so lots of your hot blood is flowing near the surface of your body. When you get out of the bath your skin suddenly starts acting like a large hot radiator and is cooled down by the air in the bathroom.

MORE ABOUT LATENT HEAT

It's interesting to reflect that knowing about latent heat loss can not only help you keep warm when you get out of the bath, it can be used to keep something cold – it's the same principle. To make a nice cold drink, place a room-temperature bottle of drink inside a washing up bowl. Cover the bottle with a damp cloth. As the cloth dries up it is losing water – i.e. liquid water is evaporating and leaving the cloth to become a gas. When it does this it needs energy. It takes this energy from the bottle and the bottle cools. How do you test whether or not this really works? Don't forget to keep the cloth damp all the time.

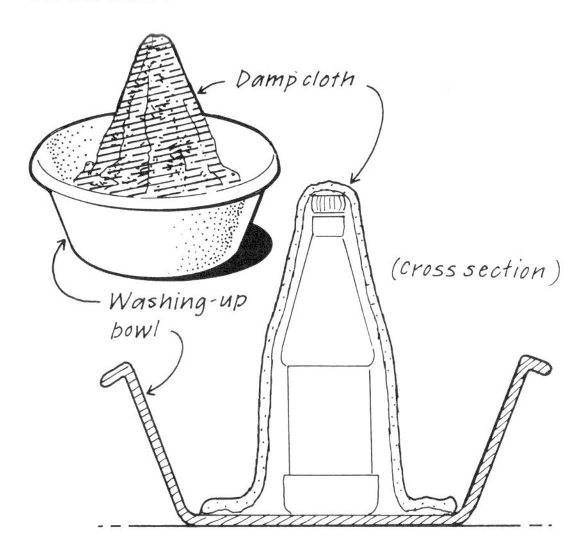

Making connections

Things which evaporate easily (some of their molecules are very light and don't need much energy to 'escape') feel cold. Try a drop of cologne or surgical spirit on the back of your hand. In addition, some understanding of all this does have some practical use. A number of things which seem tedious to young people may later make sense: like not wearing damp clothes, wrapping up after strenuous exercise, getting dry properly after showers. All of these are related to heat loss through evaporation.

Safety first

You will have your own strategies to safeguard your children's well-being in the bathroom or anywhere else. We emphasize some of the important ones.

- Hot water is dangerous – we know that – children often do not. None of the discussions in this section should lead children to feel that it is safe to play with hot water or the hot tap.
- Never encourage experimenting with any of the chemicals found in the bathroom. Toilet cleaners and the like are obvious – they are powerful corrosives and we use them to kill other living things such as harmful bacteria. Keep them where children cannot get hold of them. But avoid the danger of thinking that 'nice' things like toiletries or cosmetics are clean and wholesome in this respect. Deodorant, talcum powder and many cosmetics can be dangerous if handled incorrectly. Equally important, never mix any of the chemicals you find in the bathroom: the results are completely unpredictable and you could easily produce unpleasant or noxious gases.
- Always closely supervise games or activities that involve children being immersed in water – even shallow water – what if they should bump their heads?
- Any empty containers you use for an activity should always be clean. This may seem obvious but using empty bottles from clean sounding things like shampoo and conditioner could be harmful to children's skin and, more seriously, their eyes. Don't confuse green with safe either – they're no different in this respect.
- *Never* use containers which have previously contained bleach, even if they have been washed out.

LIGHT SWITCH

Mains electricity kills, but we live with it quite safely because we obey some simple rules. The most important is never touch it or you become part of the electrical circuit. You come closest to doing this is when you switch a light on or off. For simplicity you can assume that the wire is at about 240 V and that the floor that you are standing on is at about 0 V. If you were to touch the wire then there would be a difference of roughly 240 V across your body and you become a component in an electrical circuit, which is something you are not designed for.

What's going on?

To turn off a light you need to disconnect the wires in the circuit from each other. This is what a switch does. The switch is the only thing between you and the wires and is made out of good insulating material, usually a form of plastic. For most purposes this is perfectly safe; however, the bathroom is rather special. Firstly, the room is often filled with steam which ends up coating every surface there is. Secondly, you too are usually coated in a film of water, sometimes even standing in a puddle of it. Water is actually a good conductor of electricity especially when it has got other things dissolved in it, which it usually does have in the bathroom. It is thus good enough to conduct electricity around the plastic parts of the switch and good enough to make a good electrical connection between your finger and the water on the switch. You, the ground, the water on the switch, and the wire are all now part of an electrical circuit –

in fact you can be compared to a component in the circuit like a light bulb. Bulbs are meant to light up, however, you are not – and light bulbs don't have feelings or families.

So, this is the reason you will notice that the light switch for the bathroom is either outside the bathroom or is out of reach, usually operated by a long cord. Just because this danger occurs in the bathroom that doesn't stop it being a potential hazard anywhere else so make sure children understand that they should never touch any electrical device with wet hands.

STEAM ON THE MIRROR

CHILD: 'Look, there's grease – steam – on the mirror.'
PARENT: 'Is it always there?'
CHILD: 'No – only when there is hot water in the bath or sink – 'cos there is steam going up in the air and it goes to the mirror.'
PARENT: 'Does it go anywhere else?'
CHILD: 'Yes, up in the air and all around.'

This is part of a conversation with a six-year-old girl. We will let the educationalists argue about what kind of understanding it represents. For the authors it means she has made some connection between the misting up of the mirror, hot water in the bath and the clouds of steam rising up. More importantly it has become a genuine topic of conversation by which we can share other observations that may support her growing understanding of ideas about evaporation and condensation. Ideas like this are said to take time to develop. We as parents can make that journey more enjoyable.

Making connections

Notice how the kitchen windows get steamed up when we boil cabbage. Opening windows will help to clear them. In a car too there is condensation when we get into it and our breath condenses on the cold windscreen.

WHY BOTHER WITH A BATH MAT?

Without a bath mat children may slip over. Where else do they remember it being slippery? What other materials resemble the bath mat? What is it about the bath mat that makes it work? Contrast the effect of the bath mat with that of the soap. Can you work out what it is about a film of water that makes it a good lubricant?

Something to do

Try skimming a bottle top on a table or board top. See how much further it goes if there is a film of water on the surface.

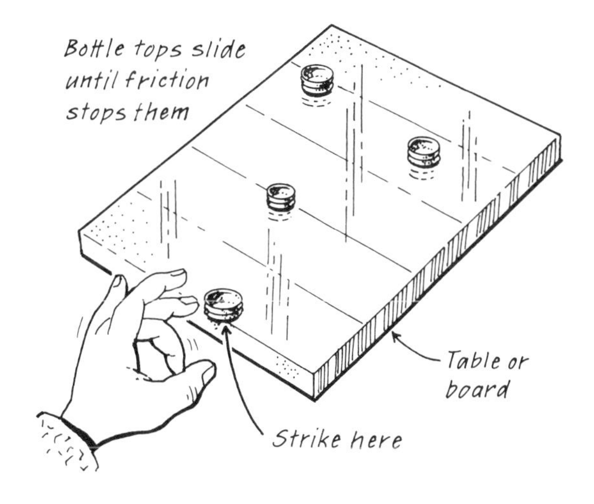

Devise a game where you try and make the bottle top come to rest in specific places after only one push. Try your own version of shove-halfpenny using some old washers. Try using dry substances like custard powder or talcum powder instead of the water to make the surface more slippery.

What's going on?

Both the water and custard powder are acting as lubricants. They get between the two sliding surfaces and reduce the friction between them. This is what oil does between moving parts in a car. There is more on friction and this type of effect in *Towns* on page 53.

PLAYING IN THE BATH

At the same time as it's good fun, playing with water allows children to gain experience of what a fluid such as water does – in other words properties such as pouring capacity and floating and sinking. It's a good idea from an early age to collect a variety of clean empty containers. Toys designed for the beach or sand pit are also good. Sand and water share some properties – perhaps you could discuss these. Ideas about properties will be treated at school through activities which are actually not much different.

FLOATING AND SINKING

Fun will always be had trying to sink toy ducks or beakers. Try and hold an empty plastic bottle under the water and see how it jumps up when you let go. How can you make it jump higher? How can you stop it jumping at all? Feel what it's like trying to push a plastic cup down in the water upside down. Feel the force of the cup trying to push up. Is there any water in the cup when you do this? Test your ideas by seeing if you can keep a ball of paper dry inside the cup when you sink it. Now, can you make a bottle float the right way up without it tipping over?

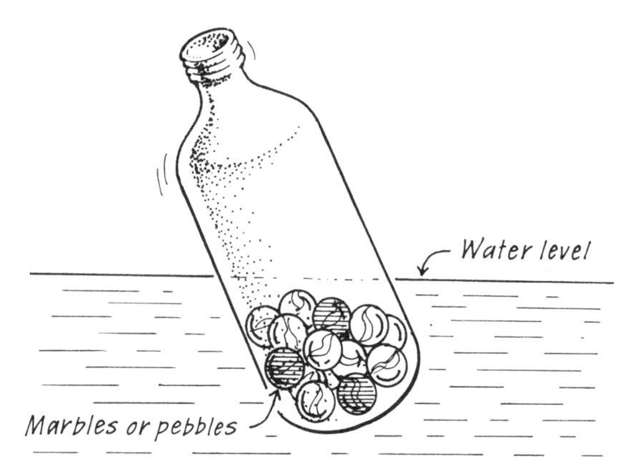

When does the sponge sink and when does it float? If you get the chance, some of these things can be even more spectacular tried out in the sea: have you tried standing on a beach ball in the water, for example, and seeing it launch when you let it go?

If you want to build on the appetite that any of the above activities have produced then all you need is a bucket of water or an outdoor plastic pool and a hot day.

Something to do

Collect a range of things you can find around the house, e.g. bottle tops, coin, pencil, lolly stick, plastic spoon, sweet wrappings, etc. Together, play a game to guess which objects will float and which will sink. Let the children feel the objects and put them in the water themselves. You may want to talk about your reasons for guessing one way or the other. How well did you do? Did you use any rules to help you? For instance, from the group of objects you happen to have a child could well start to build up the idea (say, from seeing leaves float) that perhaps all green things float.

Were there any surprises? It's the surprises that can make children think again. After a lump of green plasticine sinks they will be able to say 'it's not all the green ones after all.' Young children will be happy enough for this to be simply a game; older children might want to go further. You can even make it into a competition although some people prefer not to encourage this approach.

Things to explore

As they work through these ideas, a very simple explanation which some children put forward is 'light things float, heavy things sink'. If there is sufficient enthusiasm from an older child then it may be worthwhile testing this idea.

Make three balls of plasticine sufficiently different in size so that you can tell by feeling them that they weigh different amounts. Use ball shapes so that it's clear that they are all the same shape. Again, try and make predictions. If all three sink then this might mean plasticine always sinks, but are you sure that the smallest ball was small enough? In fact, for most primary school age children it's very good to be able to draw the general conclusion that it's the stuff that things are made of that means that something is going to either float or sink. Many children do not fully grasp this until later if they have not had the chance to test out their ideas. If you think the time is right you can also compare one of your tiny pieces of plasticine with a large piece of wood. The result might lead to a fresh line of enquiry.

National curriculum note

'How can you make the floaters sink and or the sinkers float?' Some schools use this activity to help with children's assessment at the end of key stage 1. Children can usefully go on doing this kind of thing right up to the age of sixteen, using different tactics and modifying their ideas as they go. For more, see the lemon experiment in the *Kitchens* section on page 36.

Making connections

Why do metal boats float? What difference does it make if a sweet wrapper is rolled up instead of being flattened out?

MAKING PLASTICINE FLOAT

Plasticine, like metal, always seems to sink. Take one of your balls of plasticine. Check that it sinks. Now, using your thumb, press down and turn your solid ball of plasticine into a coracle shape or if you prefer mould it into a boat shape. Whichever method you use, it's important that you neither gain nor lose any plasticine in turning your ball into a boat. If the sides of your boat are thin enough, it should float.

Once you've got your boat made between you, you can start to experiment. Investigate your boat to see what you can alter. Can you adjust it so that it 'just' floats or 'just' sinks?

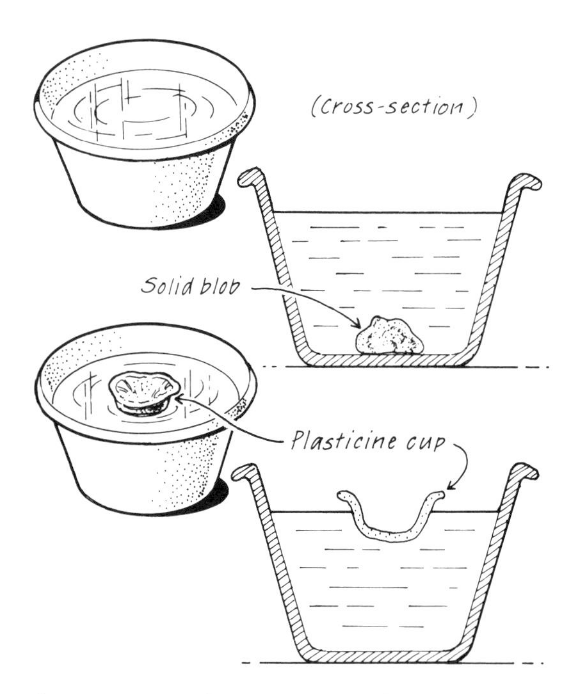

Same amount of plasticine: one floats, one sinks

What happens when you try to transport toy people in your boat?

Things to explore

For some it is enough to establish that there is something about a boat's shape that is important. Others might get the idea that it's the amount of water that you can push out of the way that's important (i.e. how deep in the water your boat sits). This is very advanced indeed. Pupils of sixteen might be able to argue in terms of their own version of Archimedes' Principle (opposite) or use advanced concepts such as relative density. Children who at eight or nine get the idea that taking up more room in the water by trapping air helps things float, are thinking very well indeed. With the plasticine experiment, although it's the same amount of plasticine each time, the boat shape takes up more room in the water – it traps air.

What's going on?

Plasticine sinks; a cork the same size floats. A solid ball of plasticine any size sinks. A cork any size floats. Plasticine is heavier than a piece of cork the same size. Plasticine is more dense (more compact) than cork; that means that plasticine is heavy for its size. A piece of lead the same size would be heavier still. It is very dense, it would definitely sink. The density of anything is how heavy something is for a certain size.

We float in water, just. Water has a density too. If you could make a life-size model person completely out of water then it would be slightly heavier than the person you had copied. The density of water is slightly greater than us. If something is less dense than water then it floats. There are things which break this rule: if you are careful you can get a needle to 'float' on the surface of clean water (don't do this in the bath). It's almost as if the water had a 'skin' on top: this is due to a property called surface tension.

Making more connections

What else floats? We do; fish don't. We use arm-bands when learning to swim. Coconuts float, that's how they get spread around the world. Is that why coconut palms seem to be found near the edge of South Sea islands? How are life-jackets made? What about buoys that boats tie up to in the harbour?

ARCHIMEDES' PRINCIPLE

If you lower an object into water then the object is going to have to push some of that water out of the way. An object will float if the weight of the water it has to push out of the way in order to sink is greater than the weight of the object itself. When you let go it will float – and the weight of water that is pushed out of the way will be the same as the weight of the object.

A final connection

If you want to float a piece of metal then you have to cheat by spreading the metal out so that it takes up more room as it enters the water. You do this by making an empty box shape and 'hiding' some air in it, which is what a boat is.

Science in the kitchen

The kitchen is one of the places in the home where parents and children most frequently interact. Children are always interested by what we are up to and quite often it happens to be the meals we are preparing for them; if not, then it's something that they're not allowed to have, in which case they are even more interested. The kitchen is a place where some mysterious almost magical changes occur. To a child, the world (unless you happen to be a dragon or a goat) often seems full of inedible things. Yet into the cooker go 'ordinary things' like flour and beaten eggs and out comes food that bears little resemblance to what went in. These changes represent important factors which children can draw on in making sense of their world.

Starting points

In this section we look at what happens in the kitchen, at food, its variety, and the changes which cooking produces. We discuss opening tins and bottles and some strange things you might notice. We talk about mixtures and solutions and try to use the scientific model of 'particles' to explain what's going on. We also introduce some optical effects a child might notice in the kitchen.

National curriculum note

The preparation of food is the focus of many traditions in a variety of cultures. These traditions are often rooted in good common sense and take account of many factors encompassing health and safety. Food hygiene is also something which we need to make children aware of.

For these and other reasons food turns out to be a popular theme for topic-based work in primary schools. Since the implementation of the national curriculum, it has been successfully used to teach children important ideas in science. The kitchen combines two virtues – it's an area rich in science and it is also an area with which children are familiar. This is always a good place from which to start developing new ideas.

Safety first

The kitchen is one of the most intriguing places in the house but it is also the most hazardous. Never leave a young child unsupervised in the kitchen. If your kitchen is very cramped then you probably don't want a young child in there with you at all. None of the activities suggested in this section need be carried out in the kitchen itself if you feel this would be unsuitable. The kitchen can simply supply the initial stimulus to go on to explore things further. All children need is their own work space and a table to work on – you can, for example, set up a table covered with plastic sheeting or newspaper in a playroom or other suitable room or in the garden.

Have you ever noticed?

- That you need to open the windows when you are boiling vegetables.
- How some things cook very quickly while others take a long time.
- How suddenly the milk can boil over.
- What happens to milk if you leave it out on the doorstep all day.
- How some things dissolve and others don't.
- How jam pots are sometimes so hard to open.
- How much easier it is to open treacle tins with a coin.
- That door handles are always on the side opposite the hinge.
- How strange things look when viewed through water.
- That a straw looks bent where it enters a drink.

National curriculum note

You are probably familiar with many of the techniques mentioned below. They employ basic principles taught at school which most children can understand by the age of fourteen. Experiencing and talking about them as they occur can provide good starting points for exploring these principles further at any age.

OPEN UP – TREACLE TINS

We can open these using a coin or the handle of a teaspoon. This is an application of the principle of levers (see diagram). If you want to experience the effectiveness of levers then you and your child can try the experiment with a door that follows.

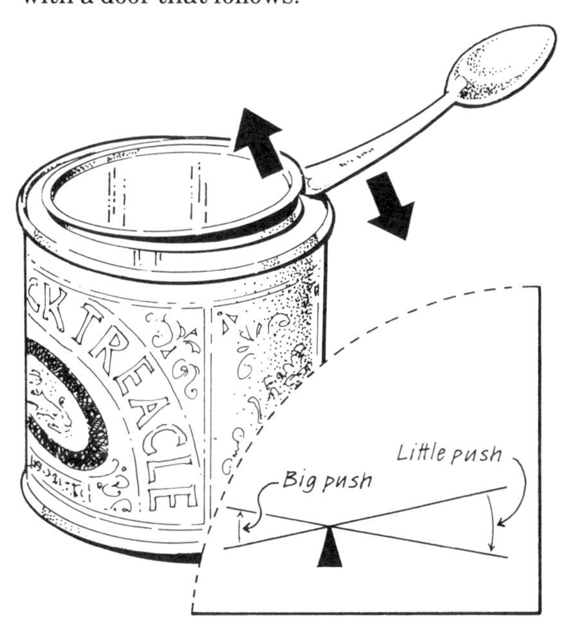

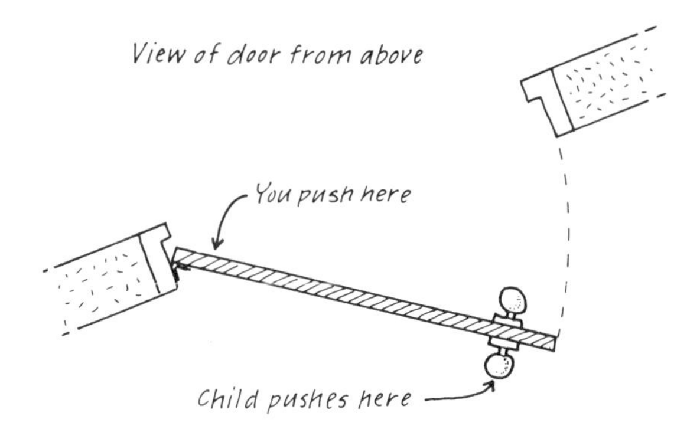

OPEN UP – JAM JARS

If it's a glass jar with a metal lid then you can try warming the lid first in your hand and it should loosen more easily. As you warm it the metal will expand more than the glass and so loosen its grip on the jar. If you think you need to, you could try warming the lid in some warm water – not hot. If you are trying this method on a plastic drinks bottle then you may be unsuccessful because the plastic will distort. Try one of the methods below.

Another way is to try and increase the friction (grip) between you and the lid. Sweaty hands are not much use nor is a layer of sticky jam. A cloth or glass paper can make the job easier. If you don't mind the marks it leaves – and most people do – you could try wedging the lid in the corner of a door.

If your hard-to-undo object is a bottle then you could resort to a pair of pliers; this is another application of the principle of levers. You may even have a gadget in the kitchen which works on this principle. For people who have any arthritic problems with their hands these gadgets are jolly useful.

Of course what lengths you go to depends on how desperate you are. I remember learning most about levers in helping my father to open a bottle of wine without a corkscrew. The technique is a family secret but a certain amount of dining cutlery was tested to destruction.

Starting points

- ▶ What do children already know about food?
- ▶ What sort of things do they say when they talk about their own meals?
- ▶ How conscious are they of the pattern of meals throughout the day?
- ▶ How do they account for the variety of diets amongst their friends?
- ▶ Do they understand that they need food for growth?
- ▶ How do they understand the seriousness of famine, or the need for a good water supply?
- ▶ Do they know that animals eat different foods than us and that we are unable to live off, say, grass or leaves?
- ▶ What do children understand by the notion of a good balanced diet?
- ▶ How do you educate your children not to eat too much tuck at school?
- ▶ At what age do you trust them to wash their hands before dealing with food without reminding them?

Something to do

When you return from a shopping trip, sort the food out together; this will teach your child something about food variety. Put away those things that need to go into the fridge or freezer straight away then you could try and sort out what's left into groups.

You can use a range of foods, e.g. packets of dried beans, tins of tomatoes, flour, pasta, etc. to play a sorting game. First try and decide what categories you would like to have and how many. Two is a good number to start with but you might find that you need more as you go on. Take turns in putting each of the foods into a category (you could start with a pile each as if they were a stack of cards). Talk about those which fit into neither of your groups and those that go into both. This is a good activity to try with children from four and upwards.

Some pairs of categories you could try include wet foods versus dry foods; food we eat for breakfast versus food we eat for dinner, or more relevant perhaps, food I like and food I don't like.

Making connections

While out shopping you can encourage children to keep a look out for the way food is grouped in shops. If you're lucky they might even help you find something. Where is the milk kept; what is it near? Why is this a different place from that of the frozen chicken? Another thing to look out for is the country of origin. With older children you could try and predict what this might be. For younger children it can raise awareness of the fact that foods do not originate in the supermarket but have actually come from somewhere else.

A BALANCED DIET

As children grow older they take more responsibility for choosing what they eat; how do they make those choices? At home, food fads can cause heartache but what effect do they have on children's health? In equipping children to make choices the most valuable help we can give is to make sure they develop the idea of variety and balance.

Young children can try and keep a record of what they have eaten over two or three days. This is more useful in reflecting variety and balance than for one day alone. At school they may make models of some of their meals and this helps give an idea of quantity as well as variety. As with the shopping try and develop categories like cooked and uncooked and discuss the focus of a meal such as fish or lentils and 'fillers' like pasta or bread. An unhelpful category to use is that of good food or bad food. Unfortunately it is one that much advertising tries to exploit. The truth is that unless you are under a special diet for medical reasons all foods are potentially both good foods and bad foods. There is no such thing as a perfect food – that is why we mix many different foods to make a meal.

With older children it is helpful to investigate the nutritional information on the packets and tins in the cupboard. Talk about how this matches any 'claims', such as 'low in fat', made by the manufacturer about the food. Survey the whole cupboard and assess it for balance. Is there anything missing that we need to obtain in other ways?

What's going on?

Our bodies need a wide variety of substances to grow and function properly. Fairly large amounts of carbohydrate (found in potatoes and pasta) for energy, more if you are going to run a marathon, less if you are going to watch it on telly. Fairly large amounts of protein to do nearly everything from moving our muscles to fighting off disease, and lots and lots of water. But none of these are any good on their own, you also need some fat, as well as vitamins and minerals. There is a vast range of these last two and although many of them are only needed in small quantities, you must have them. Vitamins are things like vitamin A, B, C, etc.; sometimes they are given chemical names like thiamin or riboflavin. Vitamins help your body do other things like get food molecules in the right places and, in the case of vitamin D, help absorb calcium into the body. Calcium is a mineral and the one which we need most of. It is important in building our bones especially when children and teenagers are growing. Most minerals are poisonous but not when eaten in the tiny quantities in which they occur naturally in food.

Sometimes people try and emphasise how each individual element in food improves some physical attribute, like carrots making you see better in the dark. True, carrots contain a chemical (beta carotene) which is used in this way but your body is unable to make use of it without a whole host of other ingredients. Not only do we need enough of the right things but we also need it in a form our digestive systems can cope with, hence the concern in recent years over the amount of fibre in our diets. Fortunately for us, this occurs naturally in foods such as apples, bananas and vegetables – otherwise we might have to eat grass like animals with different digestive systems.

Around the world, different climates and the natural vegetation have meant that people have had to develop different diets, all of which essentially accomplish the same thing. Thus the balance and variety in each one of these is very important. Each one has evolved over a period of time and some have religious significance which often helps ensure a healthy population and avoids irresponsible experimentation with potentially dangerous foods. Switching from one diet to another is not without its complications, so if your child suddenly wants to become a vegetarian then you need to look at the whole diet and seek further advice rather than simply making a straight swop between meat and beans.

HOW COOKING CHANGES THINGS

Cooking is a very sophisticated form of chemistry, the part of science that deals with the composition, structure and properties of substances and the transformations they undergo. Cooking produces changes which are the result of chemical processes. It's by studying change that scientists are able to put together a picture of the way our world works. Whenever it is safe to do so then it can be immensely valuable for children to witness some of the changes that result during cooking. Here are some which are simple to do but very rewarding in terms of the way they help children develop complex ideas.

A CHEMICAL REACTION

This is when the particles of one thing actually combine with the particles of another to form a new chemical. Here is one you can perform easily. Take a small glass of vinegar (a representative acid) and stir in a teaspoonful of bicarbonate of soda (a representative alkali) and watch. The mixture will go fizzy almost immediately.

The fizziness is caused by bubbles of the gas carbon dioxide that come from the new mixture you have just created. Once a reaction has taken place, it is very difficult to get back to where you started without some more chemical reactions and some more energy.

STRANGE TRANSFORMATIONS

Have you noticed that when you mix things like salt or sugar in water they become invisible (they dissolve) whereas things like flour or pepper don't.

What's going on?

Salt is one of those substances which dissolves in water. This can be explained by imagining that everything is made up of very small particles, too small to see. You can see a grain of salt because you've got a load of tiny particles together in a single crystal. When a grain of salt goes into water it starts to break up into these tiny particles and these are small enough to move around between the particles of water.

Eventually the particles of salt get spread out through the whole glass of water and become invisible – they form what is known as a solution. You can only 'see' salt or other solid substances when you've got a whole load of particles together. Sugar or salt that has been dissolved like this can be got back if you leave them in a saucer long enough for the water to evaporate.

When things dissolve like this you do not get a new chemical as you did with the vinegar experiment; the salt and water do not form a chemical reaction, the particles just get all mixed up. If you try and mix flour (which does *not* dissolve in water) and tap water then you get a mess. You do get particles of flour spread throughout the liquid but they are more like clumps than like the salt molecules. This is not a solution, it is a suspension. Leave it on the windowsill and see what happens. You can get the ingredients back from a suspension, usually they settle out into layers, the heaviest or most dense at the bottom.

Something to do

Allow young children the opportunity to add salt to water themselves. It may need stirring to help it dissolve. Talk about where you think the salt has gone. Comments might include 'it's gone', 'it's disappeared' or, if children have a good grasp of what's going on, 'it's hiding'. Test a range of kitchen ingredients to see which will dissolve and which will not.

At school pupils may have the opportunity to reclaim salt from solutions and grow crystals for themselves. If children persist with the idea that the salt really has gone away then you can ask them to taste a tiny bit. Of course we would normally discourage children from tasting in science experiments unless you're there to tell them it is safe.

Making connections

Have children noticed the white rings that can be seen left around puddles as they dry up? This is the salt which has been left behind as the water evaporates. They are especially noticeable when salt water pools dry up on the sand and of course, in some parts of the world, this is how some salt is obtained from the sea on a commercial scale.

A DISAPPEARING TRICK

Take a clear glass wine bottle. Put a tablespoon of salt in the bottom. Then slowly pour in tap water taking care so that the pile of salt is as undisturbed as possible. Fill the bottle so that the water level comes half-way up the neck. You now have a wine bottle full of water with a layer of salt in the bottom. Draw attention to the level of water; if you like, mark it. Now put your thumb over the top – this means that nothing can get in or out – and shake the bottle so that the salt dissolves. Look at the level of water in the bottle now. It has gone down, but nothing has come out.

What's going on?

If we like we can use this as a model to explain that when we shake it all up the salt breaks up into tiny particles which can use up the spaces left between the molecules of water. For primary school children this is worth seeing simply because it's so intriguing. In secondary schools this is sometimes used to provide evidence to support the idea that everything might be made of particles. On the other hand it could be used to show that salt really does disappear since the level did go down!

CHANGING THE PROPERTIES OF WATER

When you combine different substances the result often has different properties than those of the ingredients alone. Put a normal uncooked chicken's egg into a jug of tap water; it will sink. Children may already know that this is a way of testing whether or not an egg is fresh. What happens though if you place it in a jug containing a strong solution of salt? If you like you can set up the two liquids side by side and present it as a trick to start off an investigation into what you might have done to the second one. Once the trick has been cracked can you both think about how to devise a strength test for salt solutions?

What's going on?

Salt increases buoyancy – it's easier to float in salt water than in fresh. The amount of salt in there determines how high you (or an egg) will float. Keep adding more salt and the egg will float higher.

National curriculum note

The national curriculum for science suggests that children should have the chance to witness the forces at work in floating and sinking. This is dealt with more fully on page 24.

FINDING OUT ABOUT EGGS

Eggs are a good store of protein, they contain over 12%. They have been maligned as a cause of heart disease and food poisoning but like everything else if handled sensibly, observing good food hygiene, and with a view to variety and balance, they earn their place in our diet. If you like them that is, and if you don't there's probably a good reason.

They also perform some fascinating changes: for example, how come most things turn into a liquid when you heat them but eggs go solid? What's the difference between a scrambled egg and an omelette and what about a meringue? If you can afford it, making meringues can be a very rewarding activity for children to help with. As you make them, talk about the changes which are taking place as you whisk the egg whites and what the difference would be if you cooked the mixture quickly at a higher temperature instead of slowly at a lower temperature.

FINDING OUT ABOUT JELLY

Making an observation: you take tough jelly out of a packet, tear it up into small lumps and pour hot water over it. Then you stir it. The lumps get smaller, disappear and (if it's a red jelly) the water goes pink all over. *Questions you might share with a curious child:* why do you use hot water? What would happen if you used cold water? What about trying something in between? What things can be done to make the jelly dissolve as quickly as possible? How important is the stirring? Why might it help? Do you really need to tear up the jelly into small lumps at the beginning? What happens if you tear them up into really small pieces?

National curriculum note

Helping your child with science means allowing children to try out the answers to some of these questions for themselves where it can be done safely.

WHAT CHILDREN LEARN FROM WATCHING AND DOING

Experiments like the above lay down foundations for understanding what factors affect the speed of a chemical reaction. Witnessing the effect of temperature, stirring and the size of the pieces gives children important evidence to draw on later when they test these ideas and use them more formally. At the same time the very process of asking questions and putting forward ideas means that this simple 'kitchen sink' observation actually can give practice for essential skills in science. These will refine over a period of time and at school there will be an increasing emphasis on the ability to devise tests and make judgements.

Children will be expected to develop an increasing awareness of how fair any of their tests really are. For example, when testing the effect of using even smaller and smaller pieces, did you stir the solution by the same amount; did you even start with the same amount of jelly? There is of course no point in getting bogged down in this level of detail if it's not wanted, but it's worthwhile your listening to see how aware children are of this discipline and when they are inclined to employ it. There's also no point going beyond the stage where you start to waste too much jelly.

HEATING AND COOLING – AN EXPERIMENT WITH CHOCOLATE

Observation: a nice piece of chocolate is hard. On summer days young children will be amazed to excavate their pocket to find something rather different. Very young children might not believe that it is the same stuff. Children of any age will find the experience disappointing.

Allow children to put a piece of chocolate they feel they can spare on a saucer over a radiator or in strong sunlight and watch what happens. It would be an unusual child who has the patience to watch it all the time so you'll need to be doing something else in the meantime.

Things to talk about and try: in what ways is the chocolate different after the experiment? In what ways is it the same? What does it taste like? Which do you prefer, solid or liquid chocolate? How could you get it to return to a solid? What is the quickest way of getting it back to a solid? When it is back to a solid, is it exactly the same as before or is it different? What is the best way to keep chocolate?

What's going on?

Chocolate is one of those substances which can change from a solid to a liquid and, when cooled, change back into a solid again. Older children can begin to use this as one of their categories for grouping substances. Most of the things we do in the kitchen do not have this property. If they did then we would need to eat them before they uncooked themselves again. ▶ These are what are known as irreversible changes – an egg, once cooked and solid, cannot be made liquid again. Chocolate shares the solid–liquid property (a reversible change) with metals, some plastics and, most interesting of all, with water. ◀

ICE TO WATER, WATER TO ICE

Children of any age will want to experiment and make their own ice lollies. Five-, six- and seven-year-olds and often much older will want to make their own ice moulds from yoghurt pots or egg cartons and watch their frozen shapes melt. In this case the results should be declared inedible unless you can guarantee the cleanliness of the containers. Very young children might not automatically make the connection that ice is the same stuff as water.

FEELING COLD

Prepare some hot water, no hotter than a bath, some tepid water, some cold tap water and some ice. (Note: if this ice has come from the deep freeze then make sure it is left out for a few minutes to warm up [but not melt away]. The temperature in the deep freeze can be very cold indeed, much colder than freezing point, and can be damaging to skin.)

Take turns in closing your eyes or blindfolding each other and compare the differences between the different temperature waters by feeling with your hands. End with the ice and linger on this one for a few moments but not so that it is painful. Then, without telling them which one it is, place their hands in the cold water and ask them which one they think it is. More than likely they will say that it is the tepid or the hot. The hand is a very sensitive way of measuring but it is not accurate in telling you what the temperature is. It can only make comparisons. Having just let go of the ice the cold water will feel quite warm.

You'll find more on ice in the *Towns* section on page 53.

WITH AND WITHOUT WATER

Let children see the effect before and after leaving beans, rice or other dried food to soak overnight. Just how much bigger do the beans get? How much water do they need?

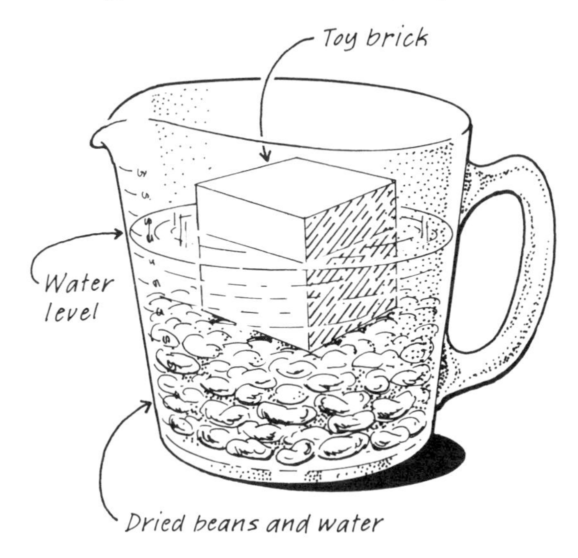

How much can the beans lift the brick?

FOOD THAT GOES BAD

It's very important to stop food from going bad – our health depends on it. For this reason, we cannot encourage experimenting with this kind of thing in the kitchen; the toxins which some bacteria produce can be lethal and a lot of moulds can give you a bad tummy ache. However, when something does go bad, it's worth drawing attention to it and talking about how you might possibly have stopped it happening. Ideas and experiments connected with decay are treated in more depth in the *Gardens* section on page 40.

What's going on?

We share our world with many other living things, including a whole host of micro-organisms, all of which are necessary to maintain balance. The air, the soil, our skin, is full of them. We cannot get rid of all of them, neither would we want to. However, some of these micro-organisms can cause disease and food poisoning. We have many tactics for minimising their effect: we can kill them as in cooking and food irradiation, or make conditions unpleasant for them so that they don't grow and multiply very well (strategies such as keeping food in the fridge or by keeping things too dry, too salty or too sweet). We can fight them with other friendly micro-organisms like those contained in yoghurt. Another way of preserving things is to keep the micro-organisms off the food by keeping it in an air-tight container but once the tin is opened, air gets in (along with airborne micro-organisms) and it's only a matter of time before that food goes 'off'. Remember it's not just the organisms themselves that are dangerous but the poisonous chemicals they produce as they eat our food; scraping off the coloured mould – say on a piece of cheese – might not mean that the food is safe.

FOOD SAFETY

It's important for children to know what is unsafe to eat and to be aware of the sorts of things involved in preserving food. The following list of ideas can be discussed as the need and opportunity arises: Know which foods need to be kept in the fridge; don't leave the lids off jam pots; know that foods need to be cooked properly; know which foods are packaged as dry foods and how to look after them; be aware of the variety of ways of processing and packaging food (tinned foods, frozen foods, vacuum packing, salting, pickling, the existence of yoghurt and similar cultured milks); know that there is the possibility of cross-infection from one food to another; know what sorts of conditions allow micro-organisms to grow and multiply.

LEMON LIFE-JACKET

Take a lemon and a mixing jug wide enough for the lemon to fit into easily. Partly fill the jug with tap water and try and predict whether the lemon will float or sink. Allow the child to feel the lemon when predicting and let them put it in themselves. Having seen that it floats, the challenge is to try and make it sink. Here's one way which you may not have thought of. Take the lemon out and remove its peel, then try it again and surprise, surprise, it sinks.

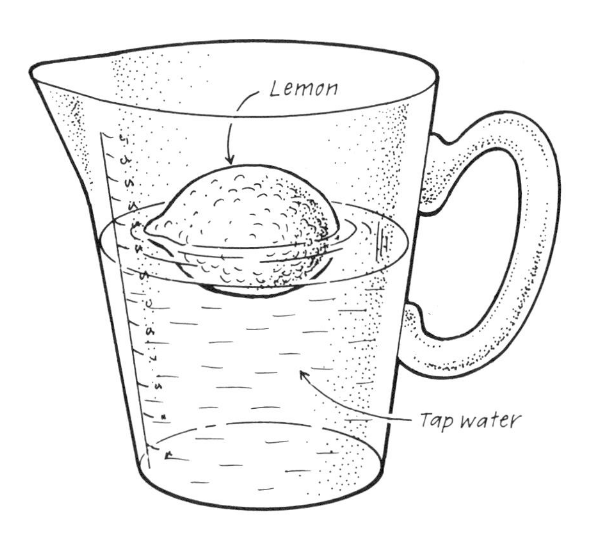

For younger children it will be enough for them to do this experiment and be amazed. You could also guess what happens to the peel alone and try that. With older children we need to be prepared to follow it up. For example, the peel on its own floats, it was acting like a kind of life-jacket, so what about looking at it more closely for some ideas? An observant child will notice that it seems to be full of holes or air pockets and if you look at the pith itself they will see that it too is holey or spongy.

It is said that this ability to float is an advantage to the fruit as it helps with the distribution of seeds and would explain why, in many Pacific islands, fruit like coconut are found near the water's edge rather than inland.

Making connections

If you want to go further then you could look at how buoyancy aids and life-jackets are made, or if you are so inclined, discuss how it fits the rules about floating and sinking you have got so far.

What's going on?

The overall density of the lemon, including its peel, is less than the density of water. If the peel is full of holes and the water doesn't fill those holes up then the lemon takes up a lot of space in the water, so it floats. Without its peel on, the lemon is more dense than the water, so it sinks. For more information on floating and sinking, see page 22.

MAKING PLAYDOUGH

When you're busy in the kitchen doing something where little hands are not wanted then this will make them feel busy too. Use a basic mix of flour and water and add salt, cooking oil and some cream of tartar to make a cohesive dough. Young children will enjoy getting the feel of this new material and mimicking you in your pastry making. It is another example of different materials being combined to produce a new material with different properties. Slightly older children will develop a sense of what these properties are and learn to take account of them as they come to terms with what you can and cannot make with the playdough. You can make good faces but you cannot make giraffes, for example.

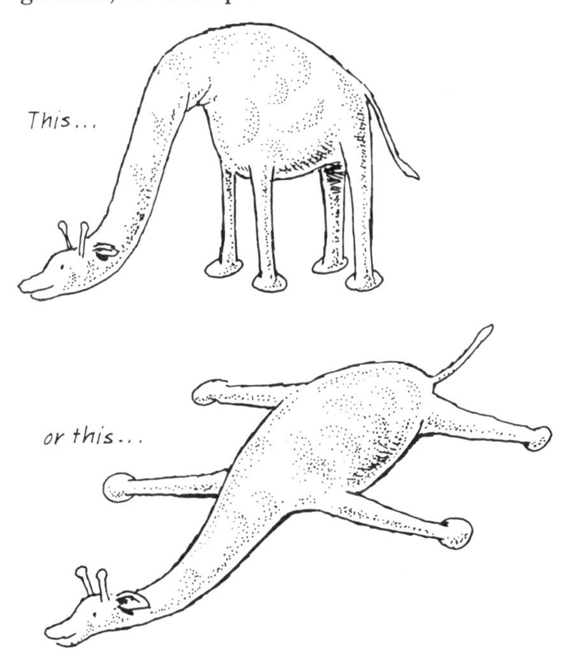

Not good for giraffes

If a child is so inclined then he or she might want to vary the ingredients to see if they can alter the properties, making it more bendy or stiff. How do you maintain these properties? What if the playdough is left uncovered?

RAISIN' RAISINS

Take a glass of fizzy mineral water or lemonade, drop in a dried raisin and watch. If you like, before you drop it in, take the opportunity to guess whether it will float or sink. Do you have any particular reasons for guessing one way or the other? Is it heavy for its size?

At first the raisin will sink to the bottom and then as if it can't manage to stay under too long it comes up to the surface again and then down again and up again and so on.

With children of any age this is worth watching just for its own sake. Older children may want to time how long it goes on for, or after looking more closely, want to try out a few ideas.

CLOSER OBSERVATION

Watch what happens when the raisin reaches the bottom. Bubbles start to attach themselves to it. More and more bubbles appear out of nowhere until there are so many that they raise the raisin to the surface. Watch what happens to the bubbles as they reach the surface. Sometimes some really big bubbles lose their grip on the raisin on the way up. They shoot off to the surface while the raisin sinks back down again. Allow young children to describe this life and death struggle in their own terms.

Bubbles form on sharp edges of the wrinkly raisin surface

Questions you could follow up with older children include looking to see where on the raisin the bubbles form. If this has got something to do with the wrinkly nature of dried raisins (our hypothesis) how can we check this idea (by carrying out an experiment).

What about other dried fruit or beans and what about other liquids? What does the drink taste like after a whole handful of raisins have been bobbing up and down for a few minutes? Can you design your own fizziness test?

What's going on?

Why are drinks fizzy? Lemonade and similar drinks contain dissolved carbon dioxide. The higher the pressure the greater the amount of this gas can be kept dissolved in it. When you open a new bottle you release the pressure that has built up, you usually hear it go 'pssht' as the gas escapes. You'll also notice that at the same time there are lines of bubbles rising to the surface. What's happened is that the drink is at room pressure which is too low for the amount of gas which is in the drink so it gets out. When you put the lid back on this gas will keep on escaping into the space above the drink. As it does this it is raising the pressure until it is high enough to stop any more gas getting out of the liquid. The fizziness you feel when you drink mineral water is the gas escaping.

Other things which affect the amount of gas you can hold in a liquid include shaking it which is what you do as you pour it out, and temperature. Children are probably already familiar with what happens when you shake a fizzy drink but what about temperature? How about devising a test to see if it's a good idea to store the drink in the fridge or not.

THE USEFULNESS OF WRINKLES (USING A MODEL TO EXPLAIN AN IDEA)

When substances dissolve, the individual particles get all mixed up together. How well they dissolve depends on things like particle size and the spaces left between them. Given the chance, molecules of carbon dioxide will club together but this is difficult when you're all spread out in a jostling crowd of water molecules. However, very sharp bends (the wrinkles on the raisin) are good meeting places where carbon dioxide can slip between the water molecules as the gaps between them get bigger and eventually form a bubble which can float to the surface. If you want to believe this (it is as yet an untested hypothesis) then you can use it to explain why the bubbles seem to form along the ridges on the wrinkly raisins. It would also explain why if you look at a glass of mineral water the bubbles always seem to be rising from the same place, probably some pimple on the side of the glass.

IS THAT ME?

Five-year-olds are amused by the sight of themselves in the back of a shiny spoon and enjoy playing 'who's got the biggest nose?' In doing this they will be altering the angle of the spoon and varying the distance they hold it away taking note of what happens as they do. Turn it over and you are upside down. Look out for other shiny surfaces: cooking pots, the surface of water droplets on the window. Once hooked you can get involved with plenty of activities using mirrors. You may have some old mirrors to experiment with, but it's safer and more interesting to use the flexible mirrors which come from school suppliers (see Useful equipment). For more ideas, see the *Toys and games* section on page 103.

Back of spoon reflection - is that me?

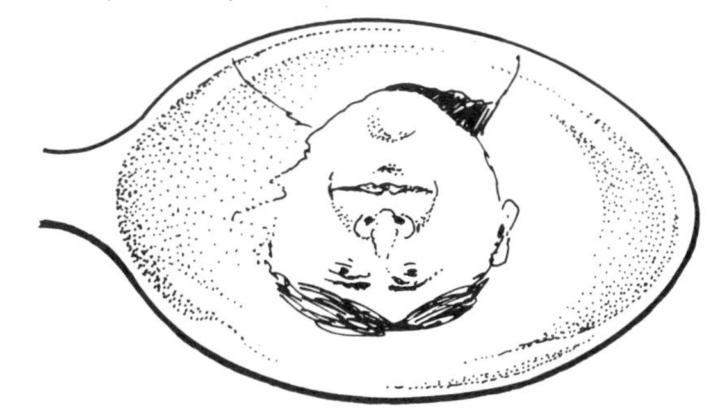

Turn it over - what do you see?

■ Science in the Garden (or Park)

If you have access to a garden you are in the position of having your very own living laboratory right outside your door. You can bring many of the things you need to look at in more detail into the house or carry out experiments in the comfort of your own home.

If you do not have a garden you can still do many of the activities described at your local park. It is worth remembering that some parks can be quite sterile places with not much in the way of 'wild' animals or plants. On the other hand many local authorities are now trying to make their parks more environmentally friendly.

All gardens are rich in science, much of it just waiting to be discovered, but the extent of that richness will depend on the kind of garden you have. A mature garden with trees and shrubs, untidy areas of stones and wood, a compost heap, the kinds of plants which are attractive to birds and insects, and a pond, all make a garden attractive to different kinds of animals. That doesn't mean to say that a more basic garden is without its possibilities. Set up a bird table and you will have an immediate resource for study. Even a small pond made from an old enamel sink will soon teem with life of different kinds. The soil and rocks in the garden are also worth investigating. Even the dustbin near the back door may well spark off ideas and further investigation.

Starting points

In this section we look at lawns, the weather, rotting and waste, growing things and minibeasts.

Note: for the uninitiated the term minibeast rapidly found favour amongst teachers in the 1970s because it overcame the problem of having to call all those creepy crawly creatures we love so much 'insects'. The problem with the term insects is that it is a name which scientists give to a single group of creepy crawlies. So it is both unscientific and somewhat confusing to refer to spiders, centipedes and woodlice as insects when they are nothing of the sort. They are in fact all invertebrates, that is, animals lacking a backbone. It is worth bearing in mind that the esteemed London Zoo still has an Insect House although it is crammed full of many different kinds of invertebrates. Strictly speaking it should be called an invertebrate house. But for young children, the word invertebrate is rather long and a bit confusing (although many of them seem to be able to cope with *Tyrannosaurus* or *Diplodocus* quite well) so minibeast has become the generally accepted term as it sounds much more exciting and friendly.

Things to notice

- At certain times the dustbins smell.
- Where does all the rubbish we throw away go to?
- Even though we cut the grass weeds seem to take over the lawn.
- Everybody says it's a good thing to put grass clippings and kitchen waste on the compost heap.
- There are lots of different kinds of animals living in the garden, particularly creepy crawly kinds.
- Sometimes it's too cold and wet to go out in the garden; at other times it's very sunny.
- Plants only grow at certain times of the year and not all plants are in flower at the same time.

GROWING PLANTS

Out in the garden or the park, as you increase your knowledge and interest, you should be able to answer the following questions. When do plants start to grow? Do all plants grow at the same time? What happens when they grow? Where do they grow? What things do they need to keep growing? What changes can you see taking place at different times of the year? Have you noticed that grass does not grow very well under trees, yet other plants seem to *prefer* a shady spot? If your child keeps a diary of events he or she can begin to appreciate the value of keeping records and note the changes which happen all around us. They can compare one year with the next. Is each year the same?

WHAT DO PLANTS NEED?

Why do we have to water plants? Do they need water to grow? Do all plants need soil to grow in? Where else apart from soil have you seen plants growing? Do plants grow in different conditions – indoors, during a drought, etc.?

You can test this idea experimentally quite easily. Remember though, in any experiment, it makes sense to do what is known as a *fair test*. This means ensuring that only one thing in the experiment changes – all the others remain the same. If, for example, you want to test what effect water has on the growth of plants try this: get three shallow containers, such as saucers or plates. Put a pad of cotton wool on each saucer, making sure they are all roughly the same size. Sprinkle approximately the same amount of seeds on each of the cotton wool pads. Place them on a windowsill which gets the sun. Cover one saucer completely with water, put enough water just to make the next one damp and leave the third one completely dry. This is a fair test. The only thing which is different is the amount of water. All the other things are roughly the same. So when you look at the results of your experiment you can be confident that it is the amount of water which is going to make a difference to the result, not anything else.

Keep a daily watch on your experiment, making sure you continue to keep the same one covered with water and the other damp. Look carefully at what is going on. Which seeds are growing? Which parts of the seeds are sprouting first, the stem or the roots? How can you tell the difference? What happens after a week? What happens after two weeks? What can you learn about watering house plants from this?

HOW DOES WATER GET INTO A PLANT?

Get hold of some sticks of celery. Cut across one piece. Look at the cut surface. What do you notice about the outside part of the stem (the side with the corrugated edge)? There are some tiny holes. They are actually the cut edges of long tubes which are the stringy bits that get caught in your teeth when you eat celery. All plants have them, it's just that they are easily noticeable in celery. To find out what they do, place a cut stick of celery in a jar of water that has been coloured with food dye. Leave it overnight and cut the end of the stem of celery next morning. Keep cutting. What do you notice about the holes now? How far has the colour gone up the stem? What does that tell you about what the tubes do in growing celery?

A variation on this experiment, which is fun to do, is to get a white carnation and carefully slit the stem lengthways into two or three. Put each cut piece of the stem in a different coloured pot of water. Leave it for a couple of days. What happens to the flower?

LOW LIFE ON THE LAWN

This activity works best in the height of summer before you cut the grass – choose a part of the lawn or the edge of the lawn where it joins the flower bed. Lie down with your child on the grass and look carefully at the life around you. Remember to have a magnifier with you as well because this can help you to get an even closer look. What can you see? What would it be like if you were the same size as a beetle in the grass? Who would be your enemies and who would be your friends? Where could you hide? How long would it take you to travel from one

Leave for a day

part of the garden to another? Where would be a safe place? Suddenly a whole new world will appear with a life all its own. With that comes an appreciation of the tremendous variety of life going on around us in a different world.

It is important for children to realise that this 'mini' world is as important in its own way as our own. Each tiny plant and animal, no matter how insignificant it seems, has its own part to play in the way the world works. Equally, as you and your child go further afield in search of scientific adventure, you can expect them to need your help in discovering the macroscopic world as well. From one of the authors' own experience (SP) it always came as a salutary experience during visits at the Natural History Museum, when pointing out the life-size model of a blue whale to five-year-olds, that all they could recognise was a large blue wall. They were unable to conceive of anything so huge as having a head and a tail since they had never had concrete experience of anything so big.

HAY FEVER

Has your child ever noticed the tall spikes of grass which stick up from the lawn in the early part of summer? With some kinds of lawn mower, they are difficult to cut. Take a closer look. Do they all look the same? With a hand lens you will be able to see they are the grass flower heads. At what time of the year do they appear? Does their appearance (and the pollen they carry) coincide with the appearance of runny noses and sneezes in the family? How long into the year do they last? What happens when you shake them on a day when it's dry and again on a wet day?

TAKEOVER BID

Once you have taken the closer look described above you might begin to ask the question, is your grass really grass – particularly if you have an untidy lawn. Although lawns which look like putting greens are almost certainly all grass, the majority of lawns are never just grass. To find out, stretch out the longest side of a wire coathanger so that it makes a roughly square shape. Drop this at random on top of the lawn and look carefully at the plants growing in it. Are all the plants in the square grass plants? You should find moss, daisies, buttercups, clover, dandelions and, if things are really bad with your garden, then occasional thistles and probably speedwell, plantain and other plants too. Make a note of the different plants and try to identify them using a plant key book (see Reading list). Keep a written record of the plants you find and try out other areas of the lawn to see if you find anything different. Why doesn't the lawn mower get rid of these plants? What do you notice about the way the leaves of a plantain grow? Do the other plants grow in the same way? How did they get there, on your lawn? Are there any bare patches on your lawn at the moment? If so, what is starting to grow there now?

What's going on?

The grasses, unlike all other plants, grow from a point level with the ground. As a result they can be cut, burnt, chewed or stood upon and still grow back. If you stopped cutting your grass, your garden would eventually turn into a small woodland. This is a phenomenon known as plant succession. Whenever a plant grows, its presence creates a new environment for a different plant which will eventually take over and so on until the patch of ground looks quite different. This can take tens or even hundreds of years. The brown patches of (dead) grass created by the dog or cat are an open invitation to new plants, their seeds blow in the wind, to take over. When you find plants other than grass such as daisies, clover and moss in your lawn, you are witnessing plant succession in action. Be reassured, however, that cutting the grass will stop the woodland.

For a longer-term experiment, try this. If you have a vegetable patch, clear a space on it, removing all plants. Sieve out the roots that are left, leaving a patch of bare soil. After a month note what plants are growing there. Keep checking each month over the summer. Where did these plants come from? Are there any that are similar nearby? If there are none, how did they get there? This activity could lead to investigations on seeds and fruits, and the way different plants use different seed dispersal methods.

HIDDEN TREASURES

Have you ever noticed the number of moths and other insects which seem to fly up at you when you are cutting the grass? They must be there all the time but you rarely notice them most of the time. The lawn is a good place to try out a hide and seek game which will give children a good idea of how important camouflage is to animals. Get hold of some different coloured drinking straws. Make sure there are some green, red and blue straws included. Cut the straws so that they are no longer than 10 cm (4 in) long. Place 20 or so on the lawn and then ask your child to find them. Time how long it takes to collect them together. Which ones were the easiest to find? Which ones the hardest? You could try this game against different backgrounds. It shows how the colours of animals help them to survive. You could test the effectiveness of different colour combinations and different patterns by painting the straws. Lollipop sticks or cocktail sticks with their sharp ends snipped off would also do.

MAKING SENSE OF MINIBEASTS

As a parent of course you may throw up your hands in horror at the prospect of getting involved with bugs you would much rather stamp on. Cast aside your concerns. Minibeasts are a major part of children's contact with the wild world and important in their understanding of what lives in it and how it works. Children have little fear of them so let them make the most of minibeasts.

Your garden will provide a rich source of minibeasts that can be used for various activities. See the section on *Pets* on page 86 for ideas for experiments you can carry out with minibeasts. First you need to be aware of what main kinds there are and how to group them. You will also need to know how and where to catch them.

Your best bet for finding them in the garden is to search around in the soil, under leaves, and under stones and logs, and to dig around in the compost heap, in the cracks in walls and fences and around the bottom of sheds.

There are so many questions you can ask about minibeasts. When you find them are there many together or are they on their own? Which ones are found together? Are they all the same kind? Where do centipedes prefer to live? When you first find minibeasts what do they do? Where do they try to go? Are the worms in the compost heap the same as the ones in the soil?

WHAT TO DO WITH YOUR CAPTURED MINIBEASTS

You can put the captured minibeasts in yoghurt pots. Use the fluffy end of a paintbrush to pick up the minibeast and do it carefully. Magnifiers and hand lenses will enable you to get a better look at them. Remember when you use a hand lens to always keep the lens close to your eye. Move the object away from or towards you, not the lens.

Here are pictures of common minibeasts and a key (see right) so that you can work out what it is you have found. Remember, after your experiment, to replace the minibeast in the area where you found it.

▶ It is much more important for a child to learn to use a simple piece of equipment such as a hand lens properly than for you to spend money on expensive equipment such as a microscope when your child is still getting to grips with the basics. ◀

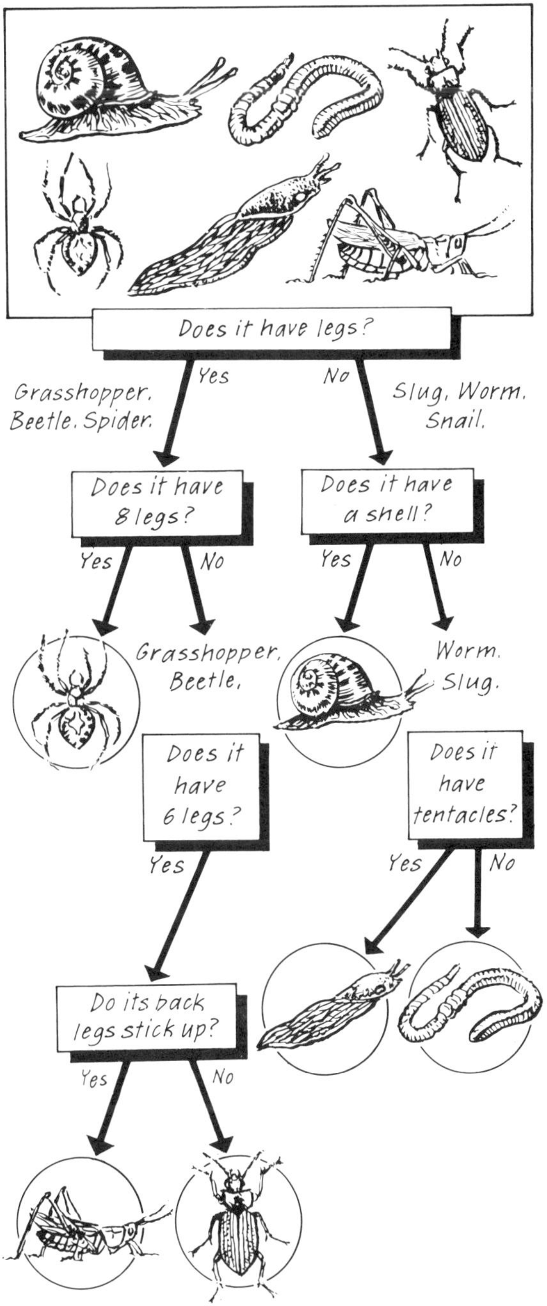

Does it have legs?
Yes
No
Grasshopper. Beetle. Spider.
Slug. Worm. Snail.
Does it have 8 legs?
Does it have a shell?
Yes
No
Yes
No
Grasshopper. Beetle.
Worm. Slug.
Does it have 6 legs?
Does it have tentacles?
Yes
Yes
No
Do its back legs stick up?
Yes
No

The dustbin or the compost heap is usually the place where rotten food from the kitchen ends up. It is important for children, whatever their age, to understand what happens to rotten food. They need to see the natural effects of rotting and be given the chance to think about this from a health and hygiene point of view, and of course to consider basic issues of environmental responsibility such as recycling.

At the simplest level, looking at the process of going rotten and then looking at the changes which take place as materials rot down is a simple experiment to carry out. This idea can then be extended to consider those materials which do not rot and at ways of recycling them. Children may have noticed the smell of a dustbin, seen the mould growing on a piece of bread, fruit or vegetable, or wondered about what happened to the waste material thrown on the compost heap.

Important ideas worth dealing with here include what other kinds of life play a part in using waste food. Natural wastes are turned into simple chemicals which can be used again by other plants and animals. Also it's important to know that not all the waste we produce rots down in this way and that some materials never rot down and can cause environmental problems. Children should also be able to appreciate that recycling can help to cut down overuse of natural resources.

Note: make sure you keep rotting material in a safe place and away from the kitchen.

Something to do – 1

Get together a selection of food materials: bread, fruit such as banana, pear or lemon, vegetables such as lettuce or tomato, and anything else you can think of. Place each piece of material in its own clear plastic bag and tie it up with a wire tie. Leave them for a week and see what happens. Some of the materials will have begun to rot. Others will have amazing moulds growing on them. It would be worth looking closely at the moulds using a magnifying glass but look through the bag rather than opening it up. When you have finished do not open any of the bags as some of the moulds can be harmful. Throw them away in the dustbin.

A variation on this experiment is to vary the conditions of the materials. For some, such as bread, add a little water. Put others in the airing cupboard or somewhere where it is warm all the time. The time of year will make a difference to the experiment. Make sure you keep records of how long all this takes. Does water make a difference? Does heat help? Will the same kind of food take the same time to rot in different conditions? All these questions can be tested and the results will help your child understand the way things rot.

Once the idea that things do rot down has been established, it will now be possible to look at the way in which some materials either do not rot or take a long time to rot. Here is where the garden comes into its own. You need to think long-term for this activity, a matter of weeks rather than days. You can turn it into a game by making estimates of if and when certain things will rot or not. Keep a note of what everyone thought and test those initial ideas against what you actually found.

Something to do – 2

Take some containers such as flower pots. Put some soil in them and add to each one some objects you think will rot and some that will not. Here is a list of some objects which you could use; anything would do but make sure there is a balance between those materials that will rot (organic) and those that won't (inorganic):

- hard vegetable such as carrot or potato
- soft fruit
- piece of newspaper
- piece of aluminium foil
- piece of clingfilm
- piece of polystyrene
- piece of biodegradable plastic bag

Water each pot as if you were watering a plant, say every three days. Leave outside. Wait a week. Using rubber gloves, empty out each pot to see what is happening to the materials that were covered with soil. How have the different things changed if at all? How much longer do they need to be left to rot down further?

Once your child has worked out the idea that rotting is a natural process which happens to some materials but not all you can begin to look for other places where rotting has or is taking place. The compost heap, piles of leaves, old logs, can all be investigated in the quest to find little rotters – those things which are responsible for the rotting down process. Break open rotten wood. What do you find inside? Try activities like sticking your hands in a pile of grass cuttings a few days after you have cut the grass. What does it feel like?

As a result of this experience the whole issue of waste and recycling might be brought up. Do we need to throw so much stuff away? Is there more we can put on the compost? What about those things which we know cannot rot down? Should you be involved in local recycling schemes such as for paper, cans, glass?

MAKING PAPER FROM WASTE

Some resources can be used again: making new paper from old, for example. This can be done by buying special kits. Alternatively, you could make it yourself by doing the following:

- tear up strips of newspaper and leave in hot water for at least two hours but preferably overnight;
- mash the paper into a pulp then add an equal amount of water;
- drain this through a mesh (nylon tights stapled or pinned to a simple wooden frame work well);
- carefully invert the mould and turn the paper sheet on to a dry towel; cover with a second towel;
- use two boards such as plywood to press out as much water as you can
- leave the paper to dry on sheets of newspaper.

You have now created new paper from old. You can use the same process to make 'cloth' paper from the woolly stuff left behind on the filter of a tumble drier.

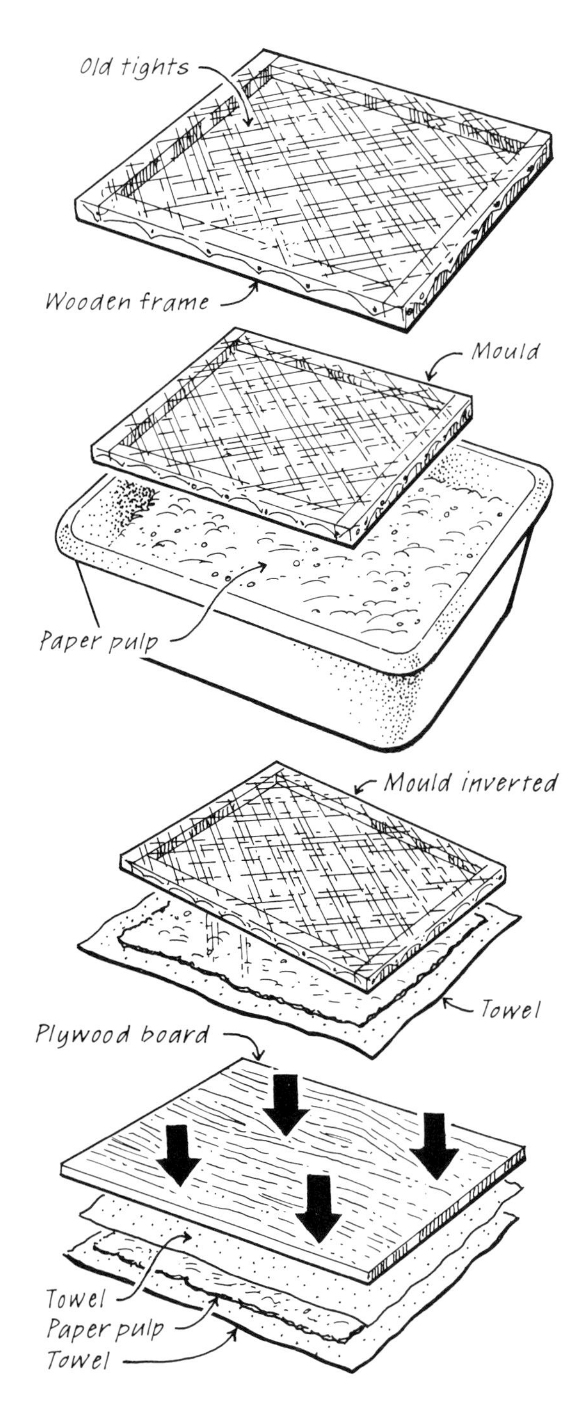

WHAT'S THE WEATHER LIKE TODAY?

The weather is one of the most important environmental phenomena that we experience on a daily basis. For children it can provide an ideal opportunity for them to make observations, be persuaded to go outside, take measurements, be persuaded to go outside, keep records and do simple calculations. Best of all it gets them outside! It can also help them watch television with more of a purpose because they can compare their results with those of the weather forecasters.

Part of the fun of studying the weather is making the equipment you need to take measurements from simple household materials. If you feel disinclined to do this, however, there are quite a few weather kits available which you can buy from suppliers or shops (see Useful equipment). But don't forget that if you make the equipment yourselves you are giving your child a really practical understanding of technology. Technology just happens to be another important subject area of the national curriculum.

MEASURING RAINFALL

Watching the weather report on TV, children may hear about so many inches of rain falling over a given period. This is a good opportunity to show them what this means. You can ask them which are likely to be the wettest times of the year, and by using a rain gauge they can check exactly how much rain has fallen. All you need is an empty polythene drinks bottle, a pair of scissors and a ruler. Cut the top off the

bottle. Turn it upside down and insert it into the cut end of the bottle to make a simple funnel. That's your gauge. Now place it in a part of the garden free from overhanging trees or bushes to avoid collecting excess water. You may need to support it with rocks or bury it partially in the ground. Every day (make sure it's the *same time* every day) remove the funnel and insert the ruler into the water that has collected in the gauge. Take the measurement, record it immediately, remove the water and set the gauge up for the next day. By keeping records in this way you are encouraging your child to be systematic in the way they collect information as well as in making observations and drawing conclusions.

WATCHING THE WEATHER

Looking at the sky and observing the clouds is an important part of encouraging children to make observations. What are the clouds like? There are three main kinds of cloud – high wispy clouds (**cirrus**); thick blankets of cloud (**stratus**), and cotton wool clouds (**cumulus**).

Looking at the wind and making judgements about what it is like is also a simple observation to make. Here is a simplified version of the Beaufort scale (which measures the force of the wind on a scale of 0 to 12) which you can use to help give an indication of wind speed. You can also make your own chart (see page 50) and fill it in daily according to the weather.

0 CALM; smoke rises vertically or almost vertically

1 LIGHT; air direction of wind shown by smoke drift

2 LIGHT BREEZE; wind felt on face; leaves rustle

3 GENTLE BREEZE; leaves and small twigs in constant motion

4 MODERATE BREEZE; raises dust and loose paper; small branches are moved

5 FRESH BREEZE; small trees in leaf sway

6 STRONG BREEZE; large branches in motion; umbrellas used with difficulty

7 NEAR GALE; whole trees in motion; walking against wind difficult

8 GALE; breaks twigs off trees

9 STRONG GALE; slight structural damage occurs (slates are removed)

10 STORM; trees uprooted; considerable structural damage

11 VIOLENT STORM; widespread damage

12 HURRICANE; whole area devastated

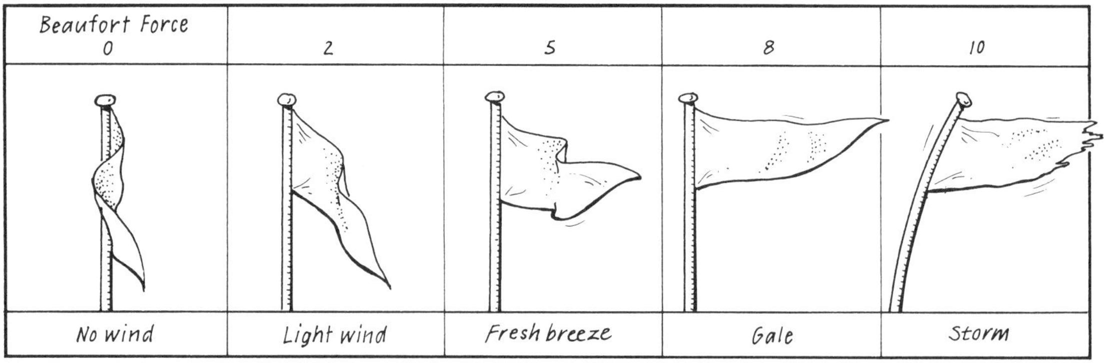

Beaufort Force 0	2	5	8	10
No wind	Light wind	Fresh breeze	Gale	Storm

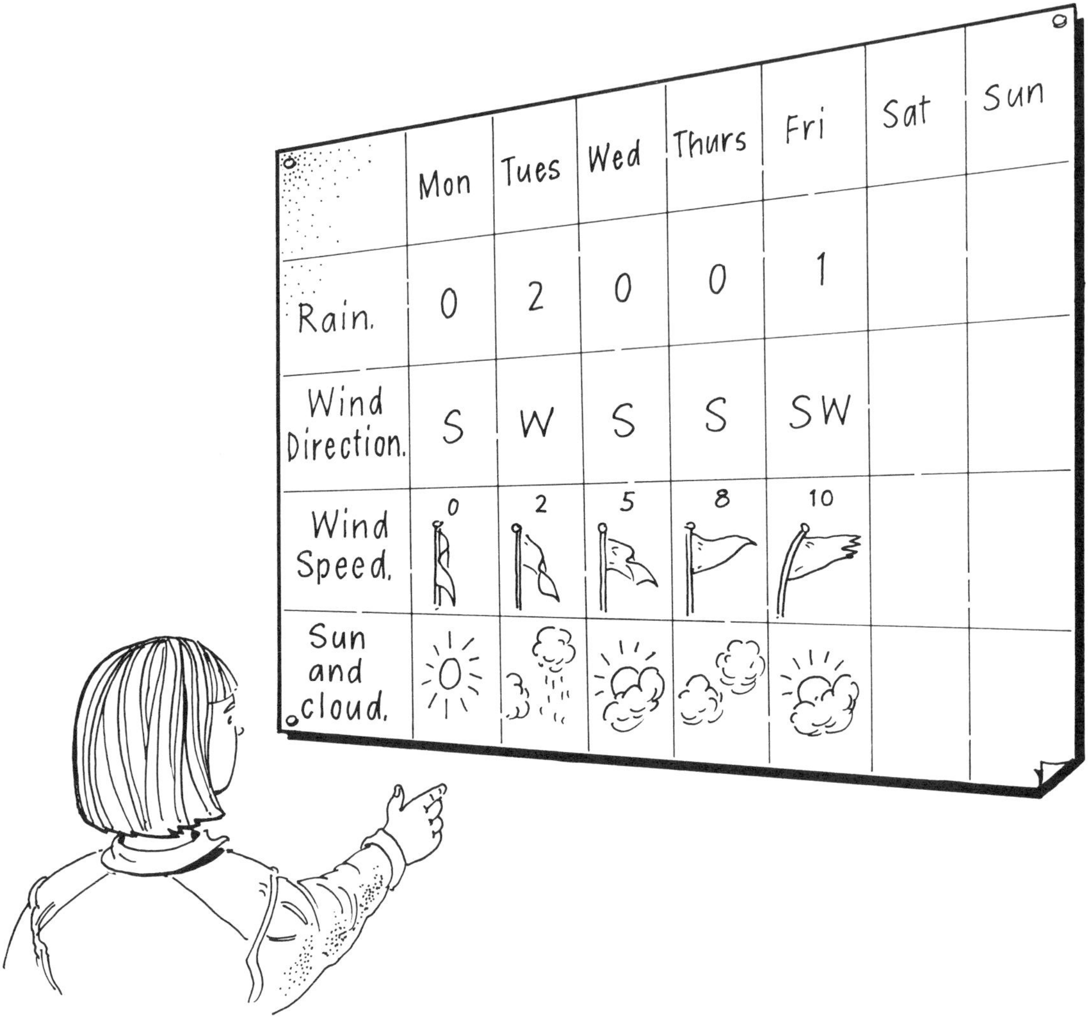

	Mon	Tues	Wed	Thurs	Fri	Sat	Sun
Rain.	0	2	0	0	1		
Wind Direction.	S	W	S	S	SW		
Wind Speed.	0	2	5	8	10		
Sun and cloud.							

CHART TO SHOW WEATHER CONDITIONS

Lay the chart out so that it can easily be filled in every day from regular daily readings. Once you have built up a number of charts together you will be able to see patterns of weather emerging.

For older children, you can build a simple anemometer (wind speed indicator) from yoghurt pots. You will need four yoghurt pots or plastic cups, the top cut from a washing up liquid bottle, a nail about 5 cm (2 in) long, a bamboo stick about one metre (3 ft) long and two long thin garden canes.

- Push the bamboo stick into the ground. Position it in an exposed part of the garden but not a wind trap as this will give you a false reading.
- Make holes in the plastic bottle top big enough to poke the sticks through.
- Push the yoghurt pots onto the ends of each of the sticks.
- Push the nail through the top of the bottle and into the top of the cane. Make sure that the whole thing spins around easily. You may need to adjust the nail accordingly.

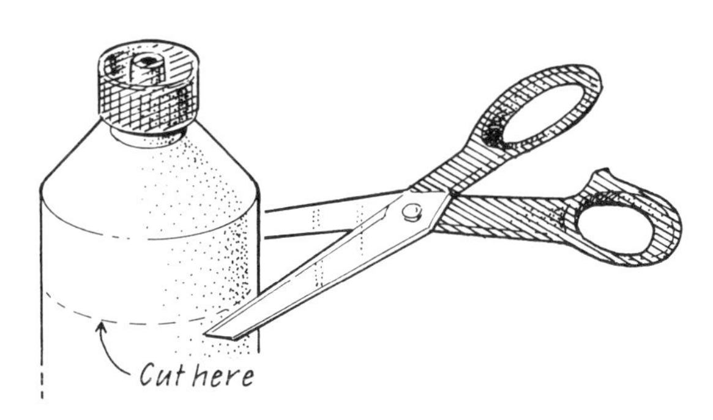

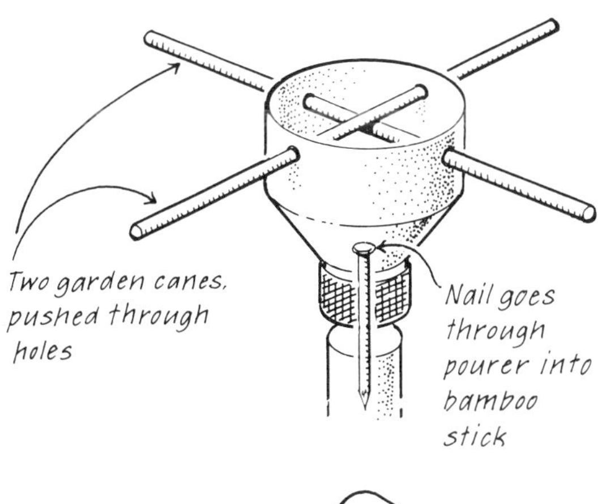

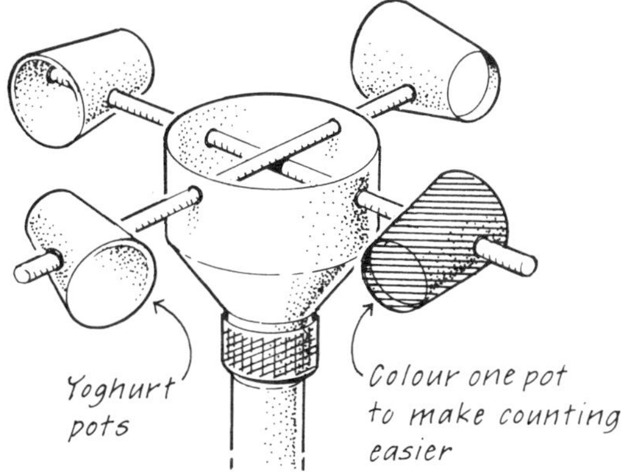

Measuring the wind speed means counting the number of times one cup spins round in a given time, say 30 seconds. Mark one of the cups with a bright colour so you can see it easily when it spins around. When you have the figure multiply by two to give you the number of turns in a minute. This figure will obviously change depending on the wind speed. You could check it against the Beaufort scale so that on a windy day the wind may be 6 on the scale and your anemometer spins really fast, and on a still day the wind is only 2 with hardly any turns on your anemometer.

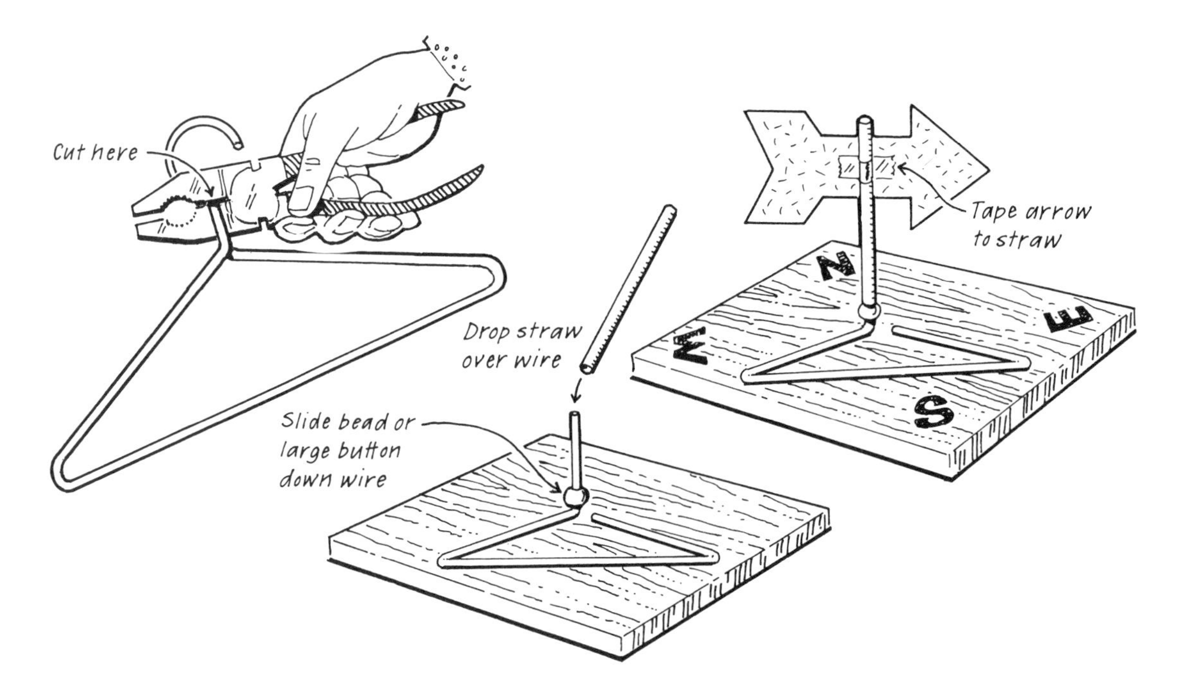

Another important thing to measure is wind direction. You will first need to know where north is so check that with a compass, or observe the position of the sun at sunrise and sunset and work it out from there. You will need a foot square block of wood, an old wire coathanger, a plastic bead or something similar, a drinking straw, and a piece of card from a cereal packet. Cut the card into an arrow shape. Stick the straw onto the arrow. Cut the bent bit of the hanger off with pliers then bend it at right angles. Pin the hanger to the block of wood using staples. Slip the bead or something similar over the wire. Drop the straw over the wire. The arrow should now spin freely. Mark the position of the compass points and you now have a wind direction finder. Remember that this device will tell you which direction the wind is blowing *from*.

Science in the town

When you're out and about it's good to be able to talk about what you see. It's also nice to keep young minds occupied that might otherwise become bored and uncooperative during long journeys or shopping trips. For pre-school children each trip out of the house is a new expedition. Older children will still find plenty to notice, but gradually the novelty wears off and without things to do trips out can become rather tedious.

When they are out and about children will have their own reasons for noticing things and without our interference will be trying to make sense of what they see. We can help by keeping alive our own curiosity and genuinely trying to make sense of things for ourselves.

Starting points

In this section we discuss opportunities that offer themselves in connection with transport, shoes, snow and ice. These are developed as examples: you will no doubt notice and want to talk about many other things. Transport is a useful topic to concentrate on. The science curriculum states that road safety is an important context in which to develop ideas about force and energy. In this section we refer to one or two examples. If there is anything we can do to help children become more safety conscious then that can't be a bad thing.

Have you ever noticed?

- The strange effects you can get from the reflections in shop windows.
- That lorries seem to have such big tyres.
- How headlights show up more easily in the dark than in the fog.
- How you tend to slip over less when you're wearing wellingtons.
- How hot and sweaty you get in a raincoat.
- How quickly roads which have been gritted and salted clear of snow compared to those everywhere else.
- Which snow is the last to go.
- What happens if you forget to oil your bicycle chain or leave your bike out in the rain all the time.
- That some buildings have sloping roofs and others don't.
- That puddles at the side of the road sometimes have rainbows in them.
- That roads wind their way up a steep hill rather than go straight up.

HOLD ON TIGHT PLEASE

For the uninitiated, there are one or two things you need to know in order to survive a journey on public transport. Children fall into this category. Telling children how important it is to hold on is nowhere near as convincing as the experience they get when the bus suddenly lurches forwards. There are few things as embarrassing as being involuntarily thrown into a stranger's lap. If you are a senior citizen then you'll understand the value of experience. It helps in being able to judge the competence of a bus driver and judge the best moment to get up from your seat before disembarking.

In secondary schools, children will be able to explain why it is that you feel like you're being thrown backwards every time the bus moves off forwards. They might do this by using terms like inertia, force and acceleration. On their way to this understanding they might describe the effect in other words. During primary school years children benefit from being able to collect their own first hand evidence.

Things to explore

- ▶ Things don't get moving unless you give them a shove.
- ▶ The heavier something is then the bigger the shove you have to give it to get it moving.
- ▶ Anything that's not tied down inside is going to get left behind.

GETTING GOING

(Something to do at home)
Take a stout glass tumbler, some stiff card and a ten pence piece and arrange them as in the diagram. Then, with the appropriate showmanship, whip the card away leaving the coin behind to fall helplessly into the glass. If you are planning to perform this at a children's party, practice first to make sure you keep the card level as you pull it out and do so at the right speed first time.

If you and your child have the inclination you can investigate this further. How slowly can you move the card and still perform the trick successfully? How many coins can you do it with? What else might you change to make the trick more spectacular?

There are also variations on this trick. One is to stack a pile of coins on the table and then with a swipe from a ruler knock out the bottom coin. Another is to use a pile of books or – if you're really brave – there's the old trick with the table cloth and teacups.

What's going on?

All of these tricks rely on the same principle to work. For example, the coin, being relatively heavy, needs a largish force to get it going and if you're going to get it going quickly you'll need a bigger force still. Your hand provides the force to get the card moving but what gets the coin moving along with it? If the coin is going to go with the card then it needs to hold on. In this case there was not enough grip between it and the card. This grip was the force that would get the coin moving. If on the other hand you had moved the card slowly then you do not need as much force. The grip between the coin and the card would have been enough to accelerate the coin away with the card.

On the bus, your feet might have a good grip but there's nothing else to push your body so the bus just drags you along by your feet until the rest of you catches up. This 'grip' force children will come to call friction.

TOY CAR TRANSPORT

When you're playing with young children and their toy cars try piling them up with bricks or giving rides to play people and see how carefully you need to drive them in order to make sure nothing falls off. Just when do you have to be most careful? See how precariously you can load up your toy cars. Who can transport the most over a dangerous journey? Who can design the most dangerous course?

COMING TO A STOP

Once you've got going, whether by bike, bus, or toy car, then of course there is the problem of stopping whenever you want to. Between the ages of six and nine, children should be getting the idea that once an object is already moving then you're going to need a force to slow it down, stop it or get it to change direction. Good brakes stop the car but what's going to stop you? The more quickly you have to stop the greater the force needed and large forces can hurt. If it helps to use these principles to ensure that nine-year-olds argue less about wearing their seatbelts then all well and good. If you want to make the point try shaking a tomato about in a sandwich box. How similar to the real situation do you think this is? To be realistic, though, the thing that's probably going to have most influence is our own example and discipline rather than common sense. After all how many of us were already in the habit of putting our own seat belt on before it became law?

Safety first

Whether children understand the concepts or not, don't allow them to throw anything from a moving car or train. Make them wait for a train to stop before getting off. Ensure they know never to get in the way of a heavy object even if it's moving very slowly. Listen out for the bangs and scraping noises boats make as they gently bob up and down by the jetty wall; although they're moving very slowly the force between the boat and the jetty can be enough to crush a child.

Making connections

When you get back home from being out in the rain just how do you go about drying off your umbrella? If you're a rapid open and closer then every time the fabric changes direction the water droplets get left behind or carry on on their own. Similarly, do you know that the best way to replace a broom handle or a hammer handle is to bang the handle into the broom or hammer head rather than the other way about?

Something to do – 1

Challenge each other to empty a plastic bottle of water without tipping it up. There are lots of solutions you might try such as syphoning it, filling the bottle with sand or making a hole in it but what happens if you simply raise the bottle into the air and then suddenly jerk it down? With practice, you leave most of the water behind. How many jerks does it take to empty the bottle completely? Can anyone do it in one go? One thing we know for sure – someone will definitely get wet.

Something to do – 2

Take an egg and, laying it on its side, spin it. Then, while it is spinning, gently but sharply place your finger on top of the egg to bring it to a halt. Once the egg has stopped remove your finger and watch. To your child's surprise, the egg will start spinning again. If it doesn't then try removing your finger a little sooner. What difference does it make how long you keep your finger on it? How do you explain what's going on? Is there a chick inside it? Maybe it gets giddy when you spin it and when you stop the shell it keeps going round inside for a bit. In that case what happens if you hard boil it?

What's going on?

When you spin a raw egg you usually spin it all, but if you stop it suddenly then you only stop the shell, the contents inside carry on spinning. The grip between the contents and the shell is not big enough to stop them immediately. If you let go fairly soon after the shell has stopped then the contents which are still spinning, gradually pull the shell round with them. If this is the case then you would not expect the same effect with a hardboiled egg. Try it.

SOME PRACTICAL STEPS FROM ALL THIS

Try and make children share responsibility for the back shelf of the car if you have one. Get them to check what's been left lying on it; books, dollies, pencil case? All of these are potentially lethal projectiles. If the car stops suddenly, what's going to stop them? If you're unlucky, the back of your head.

GOING BY BICYCLE

If you have bicycles in the family then these provide excellent talking points, especially if you trying to encourage children to take a more active part in looking after them – after all, like training shoes, they are not cheap.

- Do children understand the importance of good tyres?
- Do they check them before riding?
- How do you test if a tyre has got enough air in it?
- How many pumping actions does it take to completely inflate a tyre from completely flat? (Notice how it gets harder to pump the longer you pump for.)
- Why doesn't the air come straight out again?
- How does the bicycle pump work?
- What happens when you leave your bicycle out in the rain?

- Why do you need to oil your chain?
- Why shouldn't you oil your brake blocks?

By the time they are ten or eleven some children are lucky enough to own bikes with gears. These can be the start of a whole new range of discoveries.

- How do your legs feel if you choose to start off in top gear?
- How do your legs feel if you're pedalling along a flat road on bottom gear?
- How far does the back wheel go round for each turn of the pedals?

If children are really enthusiastic then they might get as far as changing their own ball-bearings. Even if they're not, here's a demonstration of how rolling is better than sliding as a way of getting along.

Something to do

Take a baked bean tin and try and spin it on its bottom. Now cut a piece of card to make a collar for around the bottom. Make it so that it's not quite deep enough to take a marble, then fill the collar with marbles, put it back on the table, and try again. How long can you keep it spinning for? Which tins spin the longest? Can you adapt the method for use with anything else?

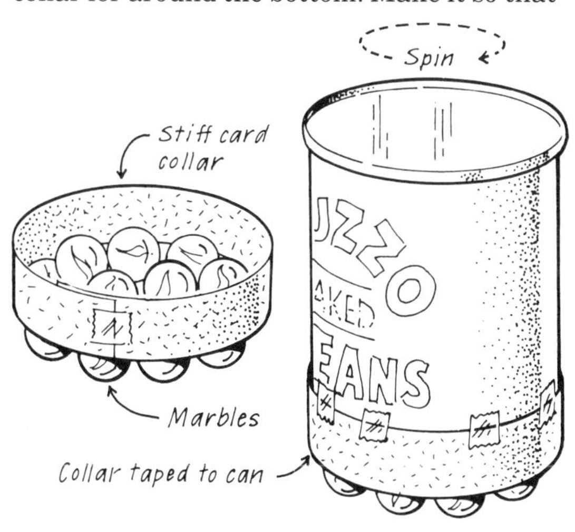

BUILDINGS AND SHOPPING EXPEDITIONS

When you go out and about with your child you might hear comments, especially from older children, about the differences between different buildings. This is more likely to happen when you go on holiday where differences are more marked but encouraging this type of observation can be quite rewarding, even in your own home town.

With young children it might be interesting enough just to notice differences in shape and style or to know whether a certain building is made of brick or stone. Your child can build on these early observations, making it easier later on to relate a building's shape to what it is made of and how old it is. As children progress they will be able to make more and more systematic comparisons between, say, a cathedral and a hospital.

TOY BRICK BUILDINGS

By playing with toy bricks, modelling clay, sand, and other materials, children get a sense of what is possible and what is not possible. Giving this play a sense of purpose can make it more rewarding, e.g. can you make a chair out of building blocks that is strong enough for the biggest dolly to sit on?

Can you make model towns which are strong enough to stand up to earthquakes and hurricanes? (You then make the earthquakes and hurricanes.)

If you want to go on further to see what it is that makes a material or a shape strong or stable then you've got to try and devise some standard tests between you but make sure your younger partner is keen first. Try not to confuse young children with too many factors at once – for example, try either to build a strong tower or a tall tower not the tallest and strongest. The best thing is to let the child lead and not to forget the fun element.

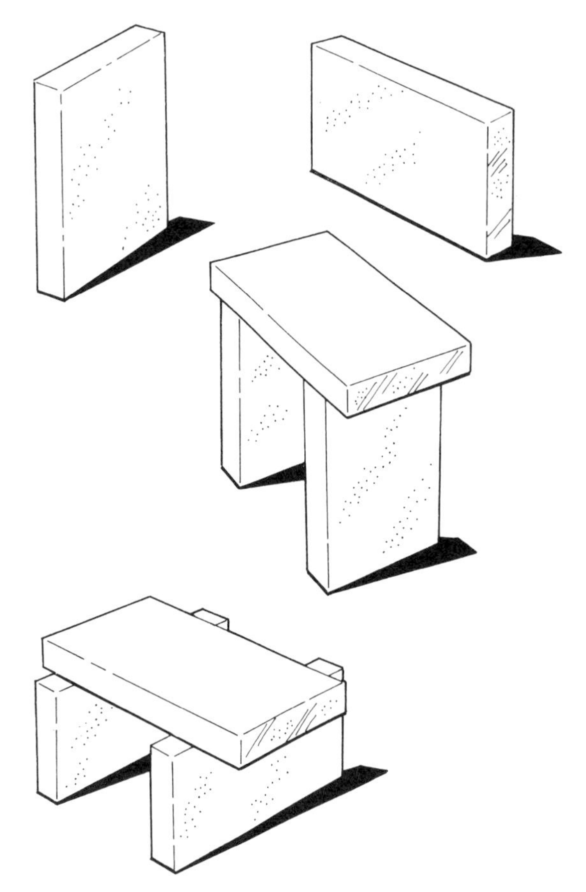

Which are the most stable (i.e. which are hardest to knock down?)

BRIDGING A GAP

A bridge is built to span either a depression (caused by a river, for example) or some obstacle, so that a road, pathway or railway can run straight across. What bridges can you and your child find in your town? What materials are used to build them? Compare an old humpback bridge made of stones or a bridge across a river built from a series of stone arches (used to span very big gaps) and a modern motorway or railway bridge.

What's going on?

Stone is strong in compression (this means that you can build stones into high piles as tall as a cathedral and they will not break under the weight of the stones above them. But stone is weak in tension: it is not capable of much 'stretch'. For buildings, this means that you have to use lots of pillars or buttresses as a support (see opposite) or build arches so that the weight is spread evenly around the curve of the arch. Many bridges use arches. Steel on the other hand is strong in tension and steel girders can be used in motorway and suspension bridges.

Something to do

Let your child experiment to find out which materials will span the biggest gap. Set up your 'span' – all you need are two level surfaces with a gap in the middle. Try a gap of 15 cm, then one of 30 cm. As the 'bridge', try using paper. Will a sheet of it do? If not, can your child make it stronger? Try folding it lengthwise to make an upside-down V-shape. The try a U-shape to make an arch.

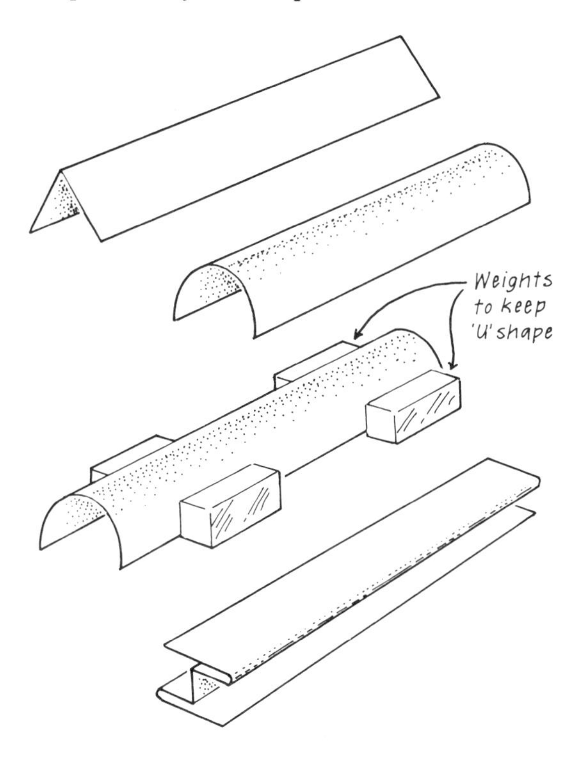

How is it possible to get a U-shape to stay in shape? By applying a little glue you can also make an I-shape. These shapes are all stronger and more resistant to bending than the original paper sheet. Test the shapes you have both made; how will you stop them from moving or sagging in the middle; is there anything you can think of to help stabilise your construction like tape or weights for the ends? Go on to try other materials like card or modelling clay – or a combination.

A WALK DOWN YOUR STREET

Investigate buildings next: what can you notice? Why do modern buildings such as cinemas or supermarkets tend to have box shapes while churches and cathedrals seem to have stone 'ribs' rather like a whale? What does this tell you about the properties of the glass, steel and concrete used for the cinema or supermarket? Both of you have a look next time you are in the supermarket: where do the heavy steel beams that support the floors and roof go – see if you can see the steelwork as a new building goes up.

GET A GRIP

Children are very conscious about what they wear. Training shoes, although expensive, are strongly desired. So apart from fashion what is it that training shoes have got over other shoes? It's worth talking to children about how they feel and trying to help young children decide at what sort of times and places it is a good idea to wear them. With older children you may be able to discuss the relative merits of how they feel, the flexibility of the sole, the sort of support they give, whether they feel like they would be good for skipping, running or clambering, and whether or not they have got a good grip.

Have a look together at the shoes on the shoe rack. With younger children, try and guess what each pair is used for. With slightly older children, try and predict which has the best grip. You will then have to try and devise your own grip tests. Like all ideas in science this will have several layers of sophistication, starting with simple push tests where you compare how it feels. If you like, you could go on to devise a test slope on which you place each shoe in turn and load it up with weights. You can increase the tilt of the slope to see which will keep their grip the longest. Do you get the same result for all surfaces? Try and compare several, first starting with ones in and around the house, then making your own test slopes to try and model what it is about a surface which makes it a good gripper.

At a deeper level of sophistication you can start to separate out the different factors. Rather than simply saying this shoe is better than that one can you make generalisations – for example, whether or not chunky soles are better than smooth ones. Are there other factors which might be more important? It's worth noting that this sort of thing is sometimes done by A-level physics students as well as primary school children. Equally sophisticated, is being able to explain that it's not just simply the shoe but what it's gripping that counts. In school, children will be developing ideas about pairs of grippers and sliders and what can be done to make them grip less.

Making connections

Really good grippers can be found on the clothes we wear, for example, when you fasten your shoes or coats using Velcro. When you get a chance, take a look at the hooks and eyes. How well does Velcro grip other material? What happens when you try to grip eyes with eyes and hooks with hooks?

Something to try

Here's a way to make a pair of surfaces which if squeezed together tightly enough produce a surprising grip. Take an empty plastic storage jar with a reasonably wide shoulder or neck. Fill it with uncooked rice. Then slide in a wide palette knife or broad-bladed cake knife. Now give the jar a good jiggle to shake the rice down and top it up with more rice as it settles. Tap the jar and press down on top of the rice to really pack the rice in hard and top up again. Once you think you can get no more rice in, slowly lift the knife and see it lift with it the whole of the rice complete with jar. (Long-grain rice works best for this test.)

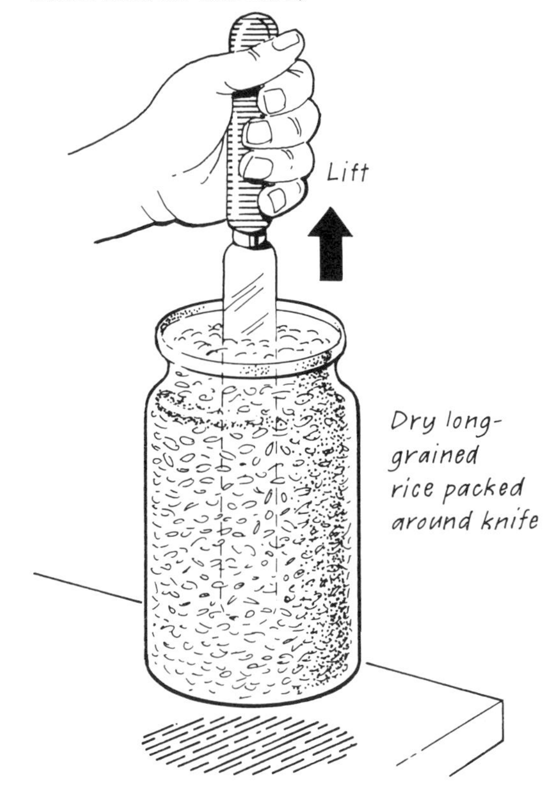

Making connections

Notice the different types of tyre on different vehicles: bicycles, cars, lorries. Why have lorries got extra tyres? Why don't all vehicles have tyres like tractors? Look at different surfaces: shop floors, paving stones, gravel. What problems are there with highly polished floors? Are all shiny floors slippery?

GETTING ALONG

Once you've got going then the next problem is keeping going. Nine- and ten-year-olds might tell you that if you don't keep pushing then friction is going to slow you down. If you intend going further than you can be thrown then you'll need to take your energy with you in the form of fuel. But whatever your understanding of force and energy it's always rewarding to be able to make a powered vehicle of your own.

What's going on?

Friction is the name we give to the force which is trying to stop objects sliding over one another. It's the force which is trying to slow everything down; you can feel it if you try and drag your hand over the table. If you press down hard you make it even more difficult to drag it. If you're driving a car then there are frictional forces between all the moving parts. If you switch off the engine then these forces will slow the car down to a halt. If it wasn't for friction then you would be able to go on for ever and ever once someone had given you a push. But since there is friction, you need to keep the engine pushing in order to overcome it.

Twenty years ago boys' annuals and some home experiment books used to say how wonderful it was to make cotton reel crawlers. Modern cotton reels, being plastic, tend not to have the same grip as wooden ones. Instead of a cotton reel you could make a crawler using a polythene drinks bottle or a cocoa tin (see page 62). These activities will need quite a bit of adult intervention as some of it is a bit fiddly. If you're using a drinks bottle you might get on better with a strip of inner tubing rather than a rubber band. It's working well when you get the elastic band to unwind slowly and smoothly rather than in jerks. When it goes along at a steady speed then the force provided by the elastic band is just equal to the forces of friction. Talk about where you think these are. Again, the science really starts when you try and improve these simple models. Of course what you have here is actually an engine. What else might you be able to do with it (see the *Toys and games* section, page 103).

Of course, today, we would hope these kind of activities are not simply confined to boys' books.

Safety first

Rubber bands can be dangerous. If a rubber band breaks it might give a flick on the hand which we can laugh off or it might flick into someone's eye which we cannot. It's worth noting that in schools if children are working closely with stretched rubber bands then they are expected to wear safety goggles, so take care.

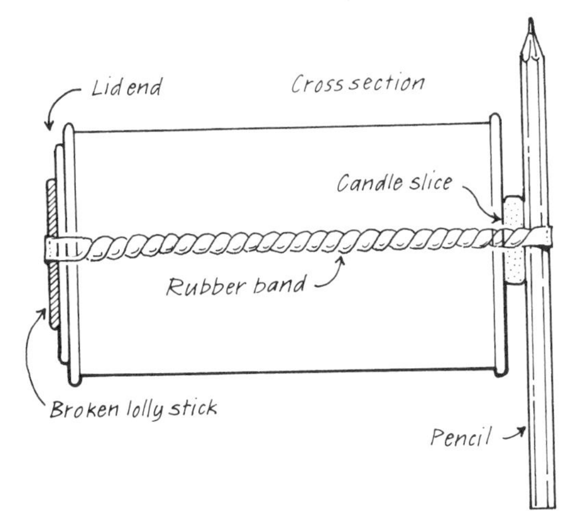

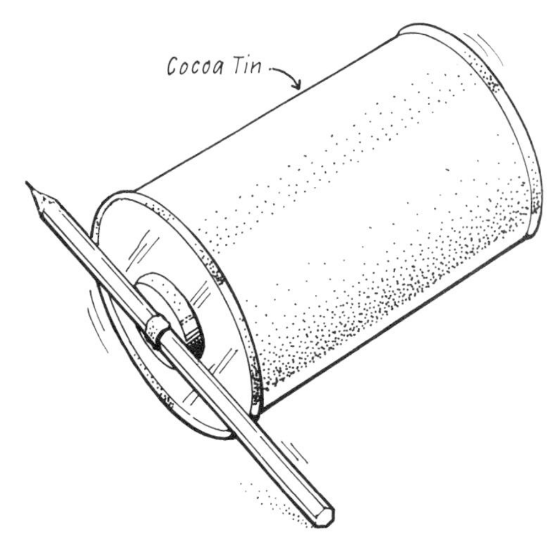

1. Punch a hole in the centre of each end (a hammer and nail works well)
2. Feed rubber band through tin and out through end.
3. Feed through centre of candle slice (candle acts as a clutch; gives controlled speed)
4. Feed pencil through loop in rubber band.
5. Feed other end of band through lid; replace lid and feed piece of lolly stick through loop in band; fix firmly with Sellotape.
6. Wind up rubber band using pencil

ON WATER

Have a look at the diagram to get an idea of what you might try to get your model moving well on water. Things to explore here are the shape, how low it lays in the water, and how to stop the propeller unwinding all at once.

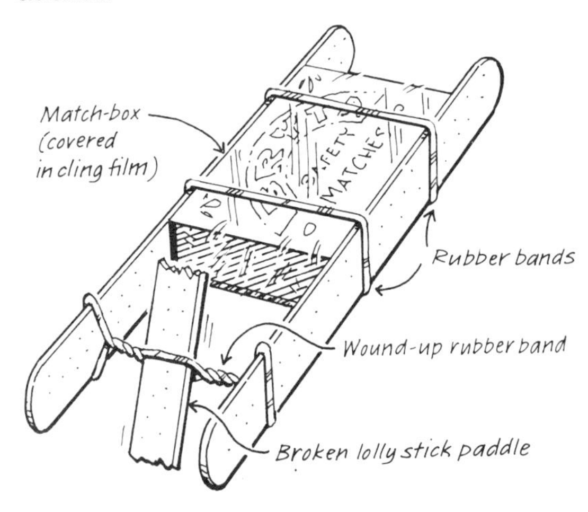

IN THE AIR

For a polythene drinks bottle crawler, you will need a model ship propeller (with a wire hook attached); you can buy these from Science Museum Exploratories or model shops. One way to get your model moving through the air (and this might be cheating) is to rig up a runway in a safe place using fishing line or garden wire. Make it slope slightly down in one direction to compensate a little for friction. Once again the fun is in adapting the idea and improving its performance. Things to think about are its size and shape together with ways of cutting down the friction between the wire hooks and the runway.

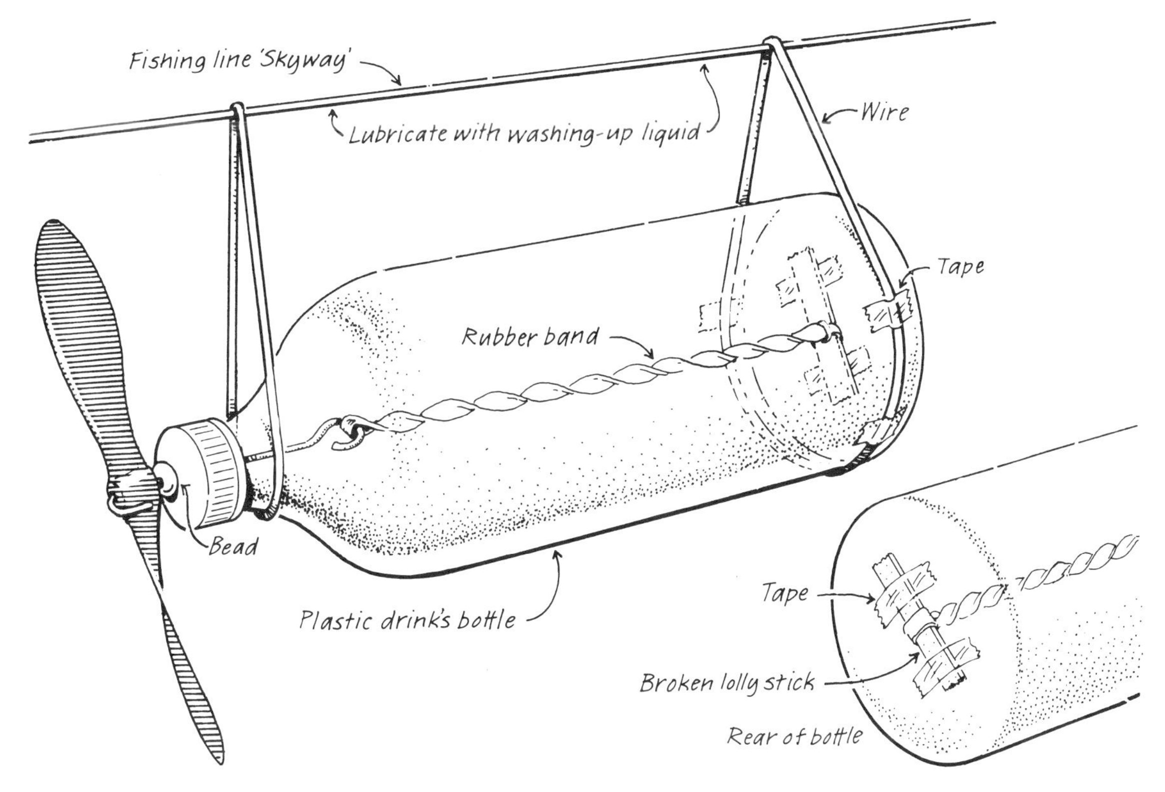

WHEN ICE FORMS AND SNOW FALLS

When there's ice and snow about outside in the garden and further afield in the town, then conditions change and there is a lot to notice. This is all the more memorable because in some parts of the UK really cold snaps happen so seldom.

Safety first

Ice is slippery. Children won't need any help in discovering this and, despite discouragement, they will always make slides. If they manage to keep them to a safe part of the playground maybe that's the best we can hope for but they should understand that slides on the pavement can be treacherous and not just to old people. Anybody can break a bone if they're taken by surprise. Sliding, though, is quite exhilarating and nine- and ten-year-olds might try and smuggle their normal shoes as well as their wellingtons and plimsoles to school because they know that they are much better sliders. But why is ice slippery at all?

Things to explore

Smoothness must have something to do with slipperiness but if you feel the ice at an ice rink you find that it is actually quite rough and there is not much similarity between this and things we know to be slippery like soap. In addition, if you look at ice skates then it might seem odd that they are so narrow; you would have thought that would have made them dig in if anything.

Something to do

Use a margarine tub as a mould to make yourself an ice brick and set it up as in the diagram then leave it. This will all take some time so make sure young minds have got something to do in the meantime rather than expect to sit and see something suddenly happen. If you return say after ten or fifteen minutes you will see that the fishing line has cut its way partly through the ice so what's going on?

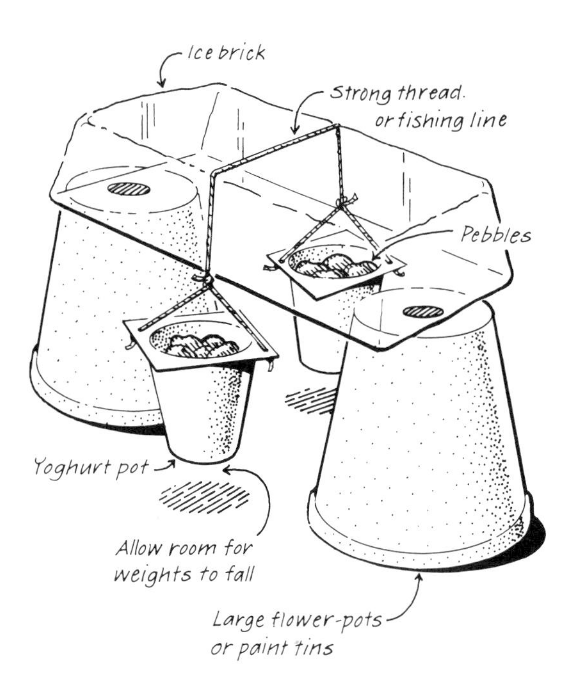

What's going on?

When you suspend the weights by the thread then you are putting all that weight in that tiny area which is in contact with the ice. This means that the ice immediately beneath the thread is under a very high pressure. Now water is one of those substances which lowers its freezing point when it's under high pressure. So if your block of ice is at minus two degrees Celsius but the melting point of ice happens to be minus four then it will melt. This then is what happens immediately beneath the thread. So naturally the thread sinks through the liquid which forms and starts to exert a pressure on the ice further down and so on. So the thread seems to cut its way through the ice.

If, however, you look at your ice there is no groove to show a cut. It looks like the thread has gone through like magic. The water through which the thread has just sunk is now above the thread. It is still very cold but it is no longer under such a high pressure, so its melting point is more probably now nearer zero degrees but the water is probably around minus two so once the thread has sunk past, the water refreezes above it. At any given moment the thread looks as if it has been mysteriously passed through a solid block of ice.

Similarly, when an ice skater skates on ice the small area of the skates creates a high pressure on the ice beneath it. Even though it doesn't get any warmer it melts so that there is a thin film of water between the skate and the ice which helps lubricate the ice. In fact it might be true to say that an ice skater skates on water.

Making connections

Sometimes when you try and make a snowman or snowballs it's completely hopeless whereas on other days it's just perfect. When you make a snowball what you're doing is to stick some of the particles of snow together. To do this you take a handful of snow and squeeze it. When you do this you are putting it under pressure and lowering the melting point so that although the snow is cold it's not cold enough to remain snow when you squeeze it. However, on really cold days you are not able to squeeze the snow hard enough to bring the melting point down to below that of the snow so that it does not stick together and feels powdery.

GETTING RID OF ICE ON ROADS

Scattering salt on ice has the same effect, that is it lowers the temperature at which the ice melts. It does not make it any hotter, if anything it actually makes it colder. Once car wheels start squashing it about too then you've got a solution of salt and water which needs to be several degrees colder to freeze.

More starting points

Snow usually takes a couple of days to melt so try and keep a record of it as it recedes. If you have got a snowman nearby then compare that with the snow elsewhere. Which will be the last snow to disappear in your neighbourhood?

SCIENCE IN THE COUNTRYSIDE

The difference between using your bathroom and the countryside for science is that the phenomena you encounter in the countryside are far less predictable and far more numerous. So numerous in fact that you cannot possibly notice everything going on around you. Unlike your bathroom which is a human-made system, built basically to pipe water in and waste out, the countryside is a 'natural' system and like all natural systems, it operates at speeds which may be too slow or far too fast for us to notice. You therefore need a guide to take you through it. Your own child can be that guide, asking questions about things you may never have thought about or even have noticed.

Starting points

What we have tried to do in this section is to indicate a few of the things which might be going on around you when you set off to explore your nearest bit of countryside. This does not necessarily mean setting off to find a spot that takes half a day to get there: the countryside can infiltrate many parts of the town: a canalside, a bit of woodland, even an urban wildlife park. Further afield, you need only look at a map to find a larger bit of woodland, well-marked footpaths or nature trails, marsh or downland, sandy or rocky coast. At the very least these can be a starting point for focusing your attention and you never know, some of the ideas outlined there might enable you to impress your child with your knowledge.

A WALK IN THE COUNTRY

Those of us with a family look to the countryside as a chance to get the kids out, somewhere for them to let off steam and run around to tire themselves out. For most children, though, particularly younger children, the countryside opens up a whole new world for them to explore. They will have rather a lot of questions about what they are experiencing, and because most of us lack the expert knowledge you need to reel off the names of plants and animals around us their questions can make us feel completely inadequate. If you find yourself in such a situation do not panic or feel you are letting your child down. Carrying such names permanently around with you is the sole domain of the enthusiast (the same enthusiast that can reel off all the names of their favourite football team and about as much use). For scientists, though, naming is a basic necessity. There is little point in studying or describing something unless you are really sure that you and everyone else knows what it is you are describing. So, to avoid confusion over regional names, names in other languages, and 'familiar' names, all scientists use a technical name which everyone agrees. The domestic dog, for example, is always *Canis familiaris*; this name is recognised internationally no matter what language you speak.

SOME 'DO'S IN THE COUNTRY

- Do encourage your children to look more closely at things, but try not to rush them. Take a magnifying glass along to help you. Encourage them to look not just at the ground but at eye level and above their heads.
- Do suggest that they sit down quietly and watch for a while – they may see things they have previously missed.
- Do look under leaves, stones, logs and behind trees to search out for small animals.
- Do make sure you do all these things together for safety's sake.
- Do encourage the use of all your senses to explore the natural world. The country is a good place to touch, smell, look and listen. It is dangerous, however, to encourage using the sense of taste. Plants and berries, even certain common ones, can be poisonous if swallowed.

AND SOME 'DON'TS'

- Don't pick flowers. Some are protected so you might be breaking the law. Never dig up any wild plant. If you do you will certainly be breaking the law, because now this is illegal. If you are keen to grow wild flowers in the garden, most garden shops stock packets of their seeds.
- Don't risk damaging or overcollecting living things if you can avoid it.
- Don't take an animal, no matter how small it is, a long way from its home. Return it to where you found it. How would you like to be dumped hundreds of miles from your home in a strange place?
- Don't leave logs and stones turned over. Always put them back in the position you found them.

Have you ever noticed?

There are so many things to notice in the countryside that it is impossible and impractical to begin to list the questions children may (and probably will) ask. The one all-encompassing question is why are there so many different kinds of plants and animals? For young children it is important that they are able to experience something of the range of this variety of life, what it is like, how it lives, and so on. The countryside is a good place to enhance that experience.

It is likely that when you go for a walk in the country you and your child will notice many things for which you have no explanation. You can, of course, always look them up later if you have the right kind of book available. To aid your memory collect pieces of material such as leaves and – for material you cannot take home – take a small sketch pad and make drawings. What we have tried to do here is to show you just a few of the things you might come across in your walk in the country. The important thing is to observe carefully and to explore. The countryside is an ideal place to encourage your child's natural curiosity.

Why do some trees have a green side to their trunk?
This is due to a kind of simple plant called *Pleurococcus* which grows as a dusty green growth on the part of the tree which faces the wettest winds. Which direction is that in your area? Take a compass with you to check. Collect some of the green dust by scraping it off and if you have a microscope look at it back home.
Why do some leaves have strange lumps and bumps on the surface?
These are leaf galls. The leaf creates a growth of leaf cells around the eggs and young of insects laid in the leaf. It does this to isolate the eggs and save itself from more damage. Each kind of leaf and insect creates a particular shape. How many different galls can you find on the leaves of trees? What signs are there that something was living inside? Try splitting one open to see what is in there.
What are the little orange blobs on the underside of bracken and ferns?
These are the reproductive parts of the plant and only appear at certain times of the year. Do all ferns have the same kinds of parts? Do they all have them there at the same time of year?
What is that frothy stuff you sometimes get at the junction of a leaf and its stem?
This is made by a leafhopper larva and is sometimes known as cuckoo spit. If you carefully move the froth to one side you will find the little green larva hiding in it. From looking at this larva can you recognise the adult form? You might see leafhoppers hopping around on the leaves of shrubby plants.
Which animal left these droppings?
Animals leave their droppings in the countryside. Each has its characteristic shape so it is possible to work out what kind of animal made it. Rabbit droppings and deer droppings are common. But which are the most common animal droppings in your part of the country? Do they belong to wild or domestic animals?
Why are some pine cones shredded?
These are the signs of squirrels having chewed the cone to get at the seeds inside. What other signs can you find of feeding in animals?

What are the little fluffy lumps you can see on or in the bark of trees or on garden sheds?
These are the egg masses of spiders. Different spider species produce different kinds. When are they most noticeable? At what time of the year do the spiders hatch out? You could collect one and keep it in a jar with air holes in the lid to see what happens.
What are those strange saucer-like objects which grow out of the trunks of trees?
These are bracket fungi which are related to mushrooms and toadstools. They grow on dead wood mostly so you can also find them on rotted tree stumps. How many different kinds can you see on your walk?
Why do some trees in a wood have white streaks down them?
These are the droppings of an owl, probably a tawny owl. Look around at the base of the tree and you might be lucky and find an owl pellet. It looks like a dropping but is actually a pellet of fur and bones thrown up by the owl. Take it home, put it in a shallow bowl of water, tease it apart and look for the bones of the owl's prey. What did it eat for dinner?
What are those long thin objects which point upwards in clumps of mosses?
These are the parts of the moss which are used to send spores into the air so they can be distributed. They act rather like a kind of pepper pot holder. When the wind blows they are shaken out. Are they there all the year?
Why have some nuts got holes in them?
Depending on what the nut looks like animals such as mice and squirrels have cut through the shell with their long pointed teeth to get at the nut inside. Look carefully and you will see the teeth marks on the nut shell.

Things to explore

Developing observation skills is an essential way to help your child and yourself to make the most of a walk in the country. Here are two games, both suitable for 5-year-olds and upwards, which you might like to try with your child. It will develop their powers of observation and knowledge of natural objects in the countryside.

MATCHING GAME

Whenever you are in the country try to spend some of the time in one place. Without your child seeing what you are doing, collect 10 objects from your surroundings. These could be leaves, stones, pieces of bark, anything in fact which is natural. Lay them out on a cloth or handkerchief and cover them so that they can't be seen. Then tell your child there are 10 objects under the cloth, all of which can be found close by. Show him or her the objects for 30 seconds and tell them to find ones just like them. When they have collected them you can compare what they have found with what you have under the cloth. Are they the same? Are they slightly different? Do you both have the same kind of leaf or are they different? This could lead to a much more focused exploration of quite a small area, and lots of spin off such as further identification you can do back at home using books (see Reading list).

COLLECTING SPREE

A game along similar lines is the collecting spree. With this game you give your child a list of things to look for. You can do this as you walk through the countryside. Some of these may be things which your child knows already, others might be unknown and yet others open to his or her own interpretation. This gives you opportunities for further exploration and discussion. The list might include the following: an acorn; a birch leaf; a pine cone; something sharp; something soft; something you like; an insect; two kinds of seeds; a thorn; a piece of human litter; something of a particular colour; a feather.

The list is endless but whatever you include will provide a focus for what you do in the country and what you do afterwards. For example, these objects can form the start of a collection of natural things. You could take the same list with you at different times of the year and see what different things you find together.

PICNIC PANIC

When you are out in the country having a picnic at the height of summer what could be worse than having your jam sandwiches attacked by marauding insects? Younger children may be merely bemused by the antics of adults flailing their hands in the air, screaming, or running away from their half-eaten food. But older children may be frightened by such over-reaction by adults and wonder 'will it sting me?' The answer is probably 'no', providing you do nothing to aggravate them. It takes nerves of steel to ignore such intruders and it helps to know what kind of beast you are dealing with. Here is some information which might be helpful the next time you are on a picnic. Perhaps this could be the time to introduce your child to the notion of warning colouration. Wasps and bees have light and dark stripes which act as a warning to other animals that they are dangerous.

ROGUES GALLERY

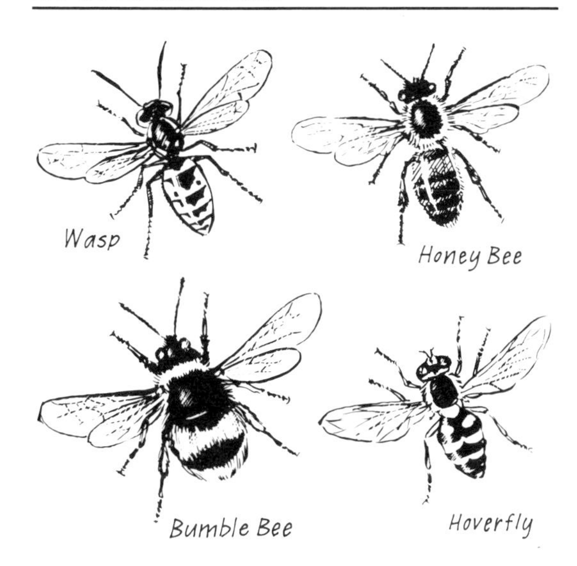

Wasp – very common; only hazardous if provoked. Can be very annoying. Important hunter and scavenger. Watch them in the late summer, particularly in rotting fruit.
Honey bee – takes some provoking before it stings. Important as a pollinator of flowers and creator of honey.
Bumble bee – very reluctant to sting. Look carefully at their bottoms to see the varied colours of the different species. How many different kinds can you find living in your area? Important for pollinating wild flowers.
Hover fly – this is the one that looks like, or mimics, a bee in order to protect itself by appearing more dangerous than it really is. It's a fly and has two wings. It hovers in front of flowers. Quite innocent and with no sting.

There are other bee mimic flies which are as large as a blue bottle. Do not be deceived. Count the wings. Flies only have two wings, not four like wasps and bees.

Once you have determined the risk factor involved you could use the opportunity, once the panic has died down, to talk about warning colours. Have they learned about being scared of striped insects whatever they are? Have you? It works doesn't it? What about the mimic flies. Would you still want to pick one up even though you know it's harmless?

Things to explore

PLANTS

Understanding the great variety of life which exists on the Earth is one of the most important ideas a child can acquire. For many children their knowledge of the animal world is quite extensive because from an early age many of the books, stories and children's characters on television are based on animals. Yet plants do not feature so highly in their experience. You can fill this missing and important aspect of science by trying out some activities around the most impressive of plants, trees.

FEEL A TREE

Try this activity to start you off. Blindfold your child or children. Lead them to a tree. Now you have to ask them to explore it using their remaining senses. Ask specific questions such as can you put your arms round it? are there any lumps or bumps on it? how does it feel when you touch the bark?

what do you smell when you get up close to the tree? what other things are living on it such as lichens or mosses? is it a living tree? is it older than you are? You might then lead them some distance away, remove the blindfold, and ask them to find their tree. Alternatively, you can explore the tree together asking your child to point out the features they remembered.

BARK RUBBING

An easy way to appreciate the variety of forms in nature is to take bark rubbings of trees. All you need is a thick piece of paper and a wax crayon. Take a leaf from the tree too. When you get home use an identification guide to work out what kind of tree it is. Take rubbings from different trees and compare them.

AGE A TREE

The most accurate way to age a tree is to count the number of rings in the cut trunk. This is not always possible although in the aftermath of a violent storm it is more likely. You do not actually need the rings because you can estimate the age from the girth of the tree trunk. It seems most trees which have a full crown (i.e. a mature tree rather than a sapling) have an average of 2.5 cm (1 in) of girth for each year of their lives. A tree with 2.5 metres (8 ft) growth would be about 100 years old. With trees growing in a wood and competing for space with others this measurement would suggest an age of 200 years. So double it when you are in the woods. Make the measurement at about 1.5 metres (5 feet) up the trunk.

SPRING BUDS

In the early part of the year collect two or three twigs from different trees. Show your child the bud on the tip. Put the twig into a jam jar of water. Leave it in a warm, light place free of draughts. Make sure your child keeps a careful watch on its progress. What they are watching is the birth of a new leaf. This process is vital to all of us. Without leaves to capture the energy from the sun, there would be no oxygen to breathe and no animal life on land.

AGEING HEDGEROWS

Wherever you go in the country you are likely to come across a hedge of some sort. Using close observation, identification and deduction, it is possible to estimate the age of a hedgerow from the number of different kinds of shrub and tree species growing there. You need not worry about correct identification. Just compare the shrubs and trees to see if they are similar or different.

Something to do

Select a 27 metre (30 yard) stretch (a metre is one adult pace or 2/3 child paces); count the number of tree and woody shrub species – include roses but ignore brambles and other climbers. Try repeating it for several stretches and average your results; avoid the ends of the hedge or gateways as these are untypical of hedges. One species of tree or shrub represents 100 years of growth. If you count five different species the hedge may have been laid 500 years ago. A whole range of historical and environmental questions are then raised. Who put the hedge there and why? Who has looked after it over the years and who cares for it now? What other things live in hedgerows?

CLEAN STREAM?

The water you drink comes mostly from the streams, rivers and ponds around us. Some of it is extracted from areas deep underground. All of it is susceptible to pollution of some sort. You can quite easily measure this pollution yourself using simple equipment. You will need to obtain pH papers to measure acidity and a nitrate kit from specialist suppliers (see Useful equipment). But if you can't be bothered with these don't worry because you can still do the fun bit without having them. This is a project where all the family can be involved. Older children will be able to grasp the ideas behind it. Younger children will enjoy finding out about the animal life to be found in a clean stream.

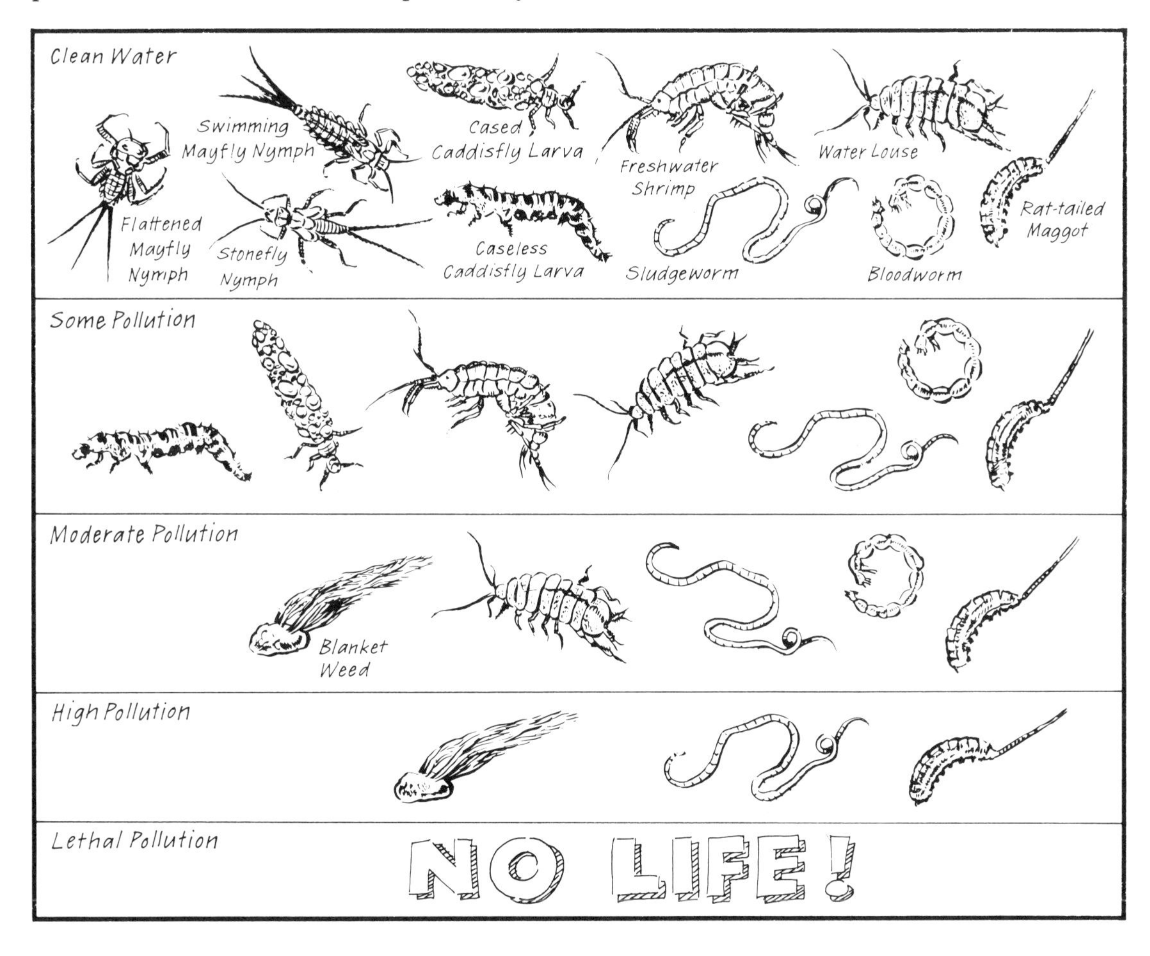

What's going on?

You are aiming to sample the animal life in a stream by collecting it in your net. By looking at the different kinds of animals there you can tell whether a stream is clean or polluted. Water authorities use this very same method for finding out about the condition of a river. These animals are said to be 'biological indicators' because certain kinds only stay around if the water is clean and there is enough oxygen for them to breathe. The chart on this page shows you those animals which indicate a clean river and those which indicate a dirty river. When pollution from an overloaded sewage works or from a factory gets into the river the bacteria which break down these pollutants multiply so much they use up all the oxygen in the water. This deprives the animal life of oxygen and it dies or is forced to move away.

What you need

- a net – you can make this from fine net curtain material or old tights stitched onto a wire coathanger, which you then strap onto a broom handle.
- a white margarine or ice cream tub, with its lid, for putting the animals into.
- a plastic spoon/widemouth pipette/artist's paintbrush for moving the animals around to get a better look.
- a hand lens or magnifying glass to get a closer look.
- some pH papers – these measure the acidity of the water (see Useful equipment).
- nitrate testing chemicals (see Useful equipment).
- rubber gloves to avoid handling the water, particularly polluted water, which can carry diseases.

Something to do

Find a stream. It needs to be fairly shallow, with a firm bed so that you can walk in it. Read the *Safety first* section on the next page. You need to get into the stream and collect the animal life living there. Do this by standing in the river facing downstream and holding the open mouth of the net in front of you and facing you. Now, gently but firmly kick the gravel and small rocks with your feet. You will dislodge many of the animals which are adapted for holding onto the rocks and gravel into the water and they will end up in your net. Take your sample and place it in the tub you have brought, making sure you have water in it to keep the animals alive. You can then see what kinds of animals there are living in the stream. Use the chart on page 73 to work out if your stream is clean or polluted.

If you have brought pH papers dip them into the stream to see what the water is like. If it's very acid it could be due to acid rain. If the pH is more than 7 (which means the water is alkaline), then there's no problem.

If you have brought a nitrates testing kit you can find out what the nitrate level is in your river. Follow the instructions in the kit. Compare the colour with the manufacturer's chart. If it is above 50 parts per million it exceeds the European Community limit. It's worth measuring your tap water to make a comparison. Don't panic if it's higher. You might like to repeat the experiment to check your reading and then contact your local water authority to find out what they have to say about it!

If you undertake this procedure on a regular basis you can see what's happening to the water in your area and whether it is being polluted.
► What you are doing is a scientific procedure for assessing the state of the environment. ◄ You are using the same methods as any scientist would and your results can be just as valid.

Safety first

Choose your site carefully. Don't go jumping into a deep stream that has a muddy bottom. Check that you are not trespassing. Wear rubber gloves at all times when handling water. Try to wash or wipe your hands after you have handled water. Avoid oversampling (removing too many animals) and disturbing the animals too much. Return all the animals you have collected to the same site. Never leave children unattended near water.

Science at the Seaside

The seaside is steeped in both pleasure and science. It is a natural place to observe, explore and discover. You don't have to work hard at helping children to be scientists here. There is so much for them to see and do. If you think back to your first experiences of the seaside what is it that you recall most of all? The warm sand, sea breeze, cliffs and rock pools, the salty sea and damp swimming costume, the hot sun and sticky ice cream. It may be different now, more crowded, perhaps even polluted, but for your children, just as novel and as rich an experience as you found it the first time around. There is so much here that we can only touch on a few ideas to get you started.

Starting points

In this section we deal with looking at tides and the moon, observing the tide line and rock pools, playing with sand and finding out about fossils.

Safety first

Safety is crucial at the seaside. Broken bottles, quick tide changes, broken sewage pipes, are all danger points. Here are some others:

- Remember that rocks are very slippery.
- Some rock pools are very deep.
- Watch out for broken glass.
- Keep away from raw sewage coming out of sewage pipes.

Have you ever noticed?

- ▶ The objects you find on the shore seem to be deposited there in a line.
- ▶ The sea is not always at the same place at the same time every day.
- ▶ Some shores are full of rocks and pebbles, others of sand.
- ▶ The length of shadows change during the day.
- ▶ Your skin can feel burnt even when it has not been very sunny.
- ▶ You cannot make good sandcastles from dry sand.
- ▶ People say the seaside air is good for you because of the ozone.
- ▶ It feels more windy at the seaside.
- ▶ Tar gets stuck to your feet or shoes no matter how careful you are.
- ▶ The water in rock pools feels much warmer than it does in the sea.
- ▶ Sea birds can sit on the surface of the sea without either getting their feathers wet or sinking.
- ▶ Ice creams always seem to melt all over your hands before you have finished eating them.
- ▶ Some cliffs along the sea have rocks in them containing fossils.

- ▶ Cliffs sometimes have layers (called strata) in them.
- ▶ There are large pipes emerging from nowhere and disappearing into the sea.
- ▶ Strange and unpleasant looking matter drifting towards you just as you are enjoying your swim.
- ▶ How quickly you want to change out of a wet swimming costume when there is a slight breeze blowing.
- ▶ How it's easier to float in the sea than in the swimming pool.

WHERE'S THE SEA GONE?

This is a common question asked by young children when they are at the seaside and can lead to some good observation and record keeping. Keep a simple tide diary every day and it will help your child see what happens to the sea and appreciate the cyclical nature of the tides.

MAKING A TIDE DIARY

You may want to explain to your child what actually happens when tides go in and out. Tides are caused by the gravitational pull of the moon and the sun. As the moon travels around the Earth it causes a bulge of water to rise on the side of the Earth facing the moon, and a similar one on the opposite side.

Time / Day	MORNING	AFTERNOON	EVENING
Mon	In	moving	out
Tue	moving	out	coming in
Wed			
Thu			
Fri			
Sat			
Sun			

The sun contributes an effect too so that when the moon and the sun are aligned they together cause an extra pull resulting in the high spring tides. When the sun and moon are at right angles to each other these pulls are cancelled out and only low neap tides occur.

Tides provide a chance to think about the moon too, so your child could keep a moon diary. What shapes does the moon have? Is it always in the same place every night? Draw its shape and position in the sky and keep your records for a period of over a month. What do you notice?

STRANGERS ON THE SHORE

A walk along the shore brings forth the sea's secrets for you to find. As you walk, encourage your child to ask questions about what they see. Here are some of the more likely things you will see with a few thoughts and questions to help you along.

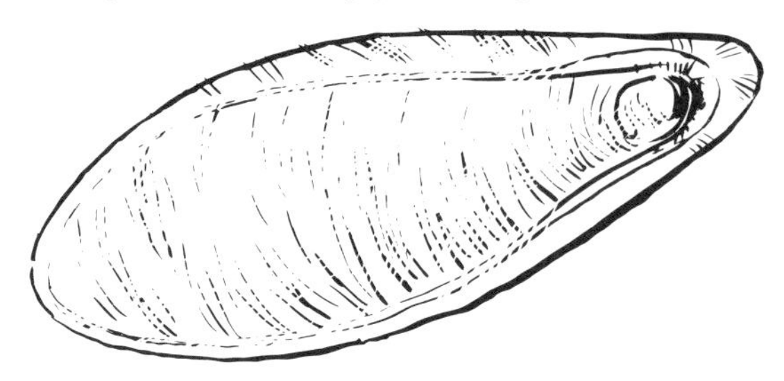

The cuttlefish bone. This is not a bone at all. Cuttlefish, like their relatives the snails (they all belong to the group of animals called the molluscs), do not have a hard skeleton inside the body. The bone is actually a flotation chamber which helps the cuttlefish stay afloat in the water. Look carefully: what do you notice about it? What's it made of? Put it in some water. Does it float or does it sink? What happened to the animal that it was part of? Take it home for the budgie.

Mermaid's purse. What could have lived in this? Some have tendrils on the four corners. Why does this help? Are there other similar ones on the shore? These are egg cases of dogfish, skates and rays, all members of the shark group. The tendrils are on the unhatched dogfish to help them attach to seaweeds. In here the fish embryo develops, feeding on the yolk inside. When it is ready, it hatches. Where is it now? Could it have ended up as fish and chips? That infamous killer fish 'Jaws' started the same way too.

Hornwrack. At first it looks like a seaweed – same shape and a similar form. But look carefully. What do you notice? Tiny honeycomb patterning is visible. Are any of the seaweeds you have found like that? When they were alive there were tiny animals living in a colony amongst the honeycomb pattern.

Whelk eggs. A large cluster of whelks hatched out of this. This is the case they left behind. How many whelks hatched out? Count the chambers.

Sea urchins. These are members of the starfish family. What do you notice about this? They are full of holes through which spines and tube feet would have stuck out. Look at a dried starfish if you find one and see what the tube feet looked like. Better still find a starfish in a rockpool. Some sea urchins like the sea potato burrow under the sand and look different from other sea urchins. If you find a sea urchin test (the name given to the hard remains of a sea urchin), count how many sections it has. At some seaside resorts you may find a sea urchin test turned into a tasteful lampshade. How many legs has a starfish? All echinoderms (the name given to animals which are similar such as sea urchins and starfish) are split into five sections.

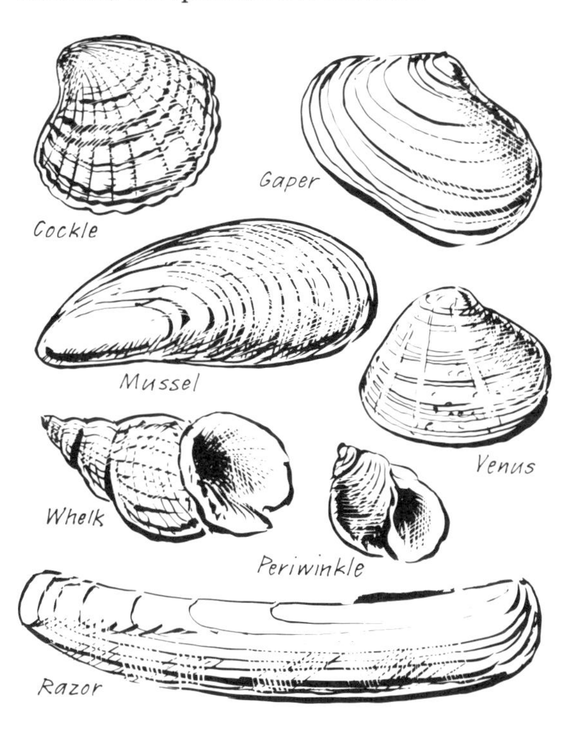

Various shells. There may be many different kinds of shell along a shore. Look at them and try to work out what kind of animal lived inside. What shape was it? How big was it? Did it live on rocks or in sand? How many parts were there to its shell? Some have two parts to the shell, an example is the razor shell. Can you find both the parts and pair them up?

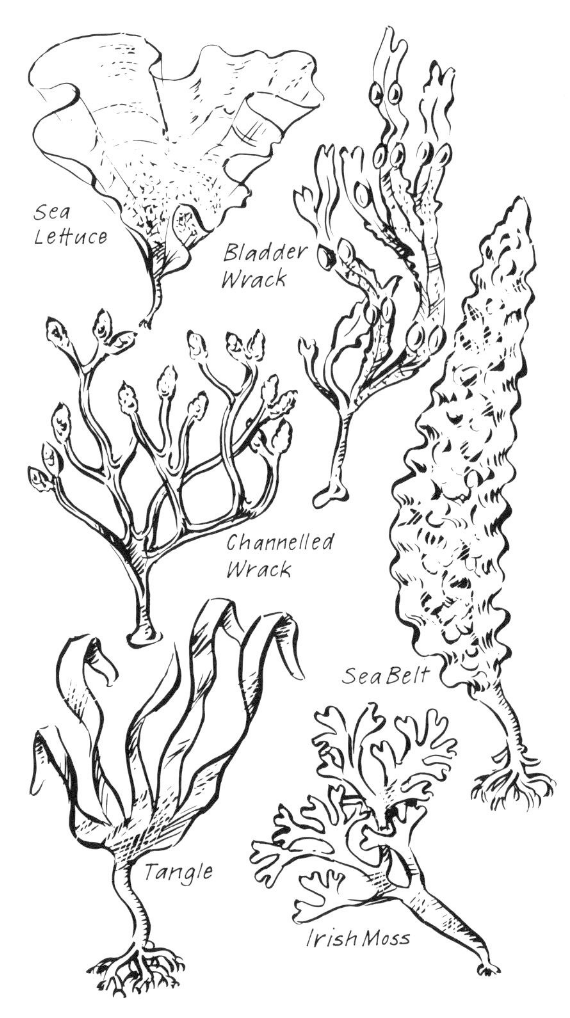

Seaweeds. There are several different kinds of these. Sort them out. Which are unbranched and which are branched? Do some have air chambers to keep them afloat? Can you see the parts of the seaweed used in reproduction? What are the similarities and differences between seaweeds and land plants? Can you find the stipe or stalk and the holdfast which a seaweed uses to grip to a rock?

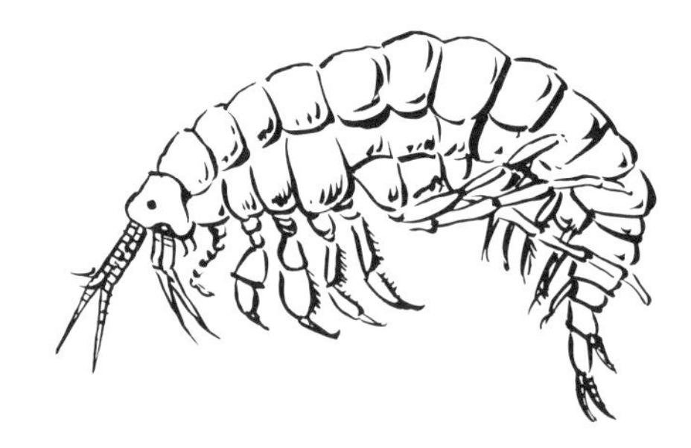

Sandhoppers. As you rummage through all this material you may notice some movement. If you look carefully you will see tiny shrimp-like animals scurrying or even hopping around. These are sandhoppers. What parts of the body are used for moving or hopping?

FOSSIL FINDING

At the seaside you will often find exposed cliff faces near or on the beach. You or your child may notice that the rocks in those cliffs are in different layers. Some of the rocks from the cliffs might also contain what look like the shapes of shelled animals. These are fossils of animals that were alive millions of years ago. Each layer tells its own story of what animals were alive in the sea at that time. The layers at the bottom of the cliff are probably older than those near the top. Not all kinds of rock contain fossils but rocks made from sand or mud may certainly contain them. Limestone made from the shells of long-dead animals will definitely contain fossils. Sandstone, mudstone and limestone are all relatively soft rocks.

If you find some fossils you might like to have your child reflect on how they resemble some of the animals you have found living on the seashore during your visit. By implication, these fossils were sea animals too. So what are they doing up the cliff? Either the sea was up there millions of years ago or the rock layers were pushed up there by forces in the Earth. Either explanation is possible. What about those layers in the rock? What made them? If there are fossils in them from the sea then the chances are they are made up of layers of sand or mud. Each layer represents a different time in the history of the part of the country you are in. For many children, particularly younger ones, the notion of going back in time is a difficult idea for them to grasp, along with the idea that animals are capable of turning into rock.

One way to get over that idea is for your child to play a game of building layers. Set up a container, say a plastic fish tank or something similar. Each day your child places a layer of material, say sand, soil, paper plus a couple of objects, in the tank. The next day the process is repeated using different objects and different materials. After a week you can dig down and relate the objects to the different days.

BACK HOME

Collect some patterned shells and take them home with you. To represent how a fossil is formed, your child can make a simple plaster cast of the shell. Simply press the shell firmly into a block of plasticine to make an accurate impression. Remove the shell and make a collar of card, fastened with a rubber band, to surround the plasticine. To make your cast, mix plaster of Paris with water until it becomes creamy. Pour it carefully into the impression up to the level of the card. Wait for it to set and then remove the card and peel the plasticine off the set plaster. You now have a 'fossil' shell.

Most fossils are in fact created not by the shell leaving an impression which is then filled by rock (as in the plaster cast), but by chemical reactions which change the shell over millions of years into rock.

SAND PLAY

When you are spending a day on a pleasant sandy beach ask your child to lie down in the sand with their swimming costume on. Ask them how it feels. How do they describe it? What's the sand like here? Is it different from sand on other beaches? Is it made just from sand? Are there other things in it?

MAKING A SAND TIMER

Cut the top off a polythene drink bottle. Invert the top into the bottom of the bottle. You now have your timer. Add some sand to the cut top and mark its position with a pen. How long does it take for the sand to fall through? What happens when you change the amount of sand? Does twice the amount of sand take twice as long as to pass through? Find a way to test this. What happens when you make the hole smaller or larger?

SAND SOUNDS

Take a couple of empty margarine containers to the seaside with you. When you are exploring, pick up different objects and place them in the containers. What sort of sounds do they make? Try sand, pebbles, dried seaweed. What does your child think is happening to the objects to make the sound?

SUNSHINE AND SHADOWS

When you are planning to be down on the beach all day here is something you can try using sand and the sun. Mark a position on the sand using a rock. Make sure you don't move it. At regular time intervals, say every hour, stand at that position. Ask your child to draw around your shadow in the sand. Maybe your child can collect some pebbles to make a more permanent record. What do they notice? Does the shape of your shadow change? What about its position? Can they explain what's going on? Can they tell the time from the shadow? What does it tell us about the way the sun moves in the sky? If you like, you could make a more permanent sundial when you get home. Many books explain how to do it (see Reading list).

BUILDING SANDCASTLES

In the process of building sandcastles children will learn about the quality of sand, what you can and cannot do with it and how water changes its consistency making it better for building. When you dig down in the sand there will be wet sand underneath and drier sand above. Nearer the sea it is wetter closer to the surface than sand higher up the beach. Ask your child why they think that is. When they build the moat around the castle they will find out about water. How long does it stay in the moat? When objects fall in the water some will float while others sink. Flags or windmills to stick in the sandcastle provide experience of wind, how it changes in strength and in which direction it blows. The process of building a sandcastle is a powerful experience to find out about a variety of different aspects of science. If you are there when it's going on you can add to your child's understanding through talking, questioning and listening.

EXPLORING ROCK POOLS

Along rocky coastlines the outgoing tide leaves behind it a temporary window into the sea – a rock pool. Here is the place where children can explore a different and unfamiliar world. Here the plants are nothing like the plants on land and some of the animals look more like plants than animals.

Have a close and careful look at the rocks around the pool and look carefully at the things living in the pool itself. What can you see on the rocks? Are there small red and pink blobs of jelly stuck fast onto the rocks? Do they bear any resemblance to the wafting tentacles of the sea anemones in the water? They are one and the same animal, even though they look more like a plant. The blobs of jelly are sea anemones which have shut up shop until the tide returns to cover them once again. What happens if you pour sea water over the blobs? Try lifting one of the blobs and putting it in the water. What happens?

Look into the pool. You may see any number of different snail-like shells stuck to the sides of the pool or littering the bottom. But some seem to move much faster than the others. Snails never moved that fast. Reach in, pick one up and examine it. Is it a snail? This is a hermit crab, a small crab which protects itself by living in an old snail shell. How many different kinds of shells do they use?

What else can you see in the pool? Crabs of different sizes? Look on the underside of some. You may find a female with hundreds of her eggs kept hidden away under her. Have a look at the crab's legs. Is there anything on the legs which might help a crab swim? Something which is a paddle shape?

You may find a lot of conical shells – limpets – stuck on the rocks. In the pool you may see them moving slowly over the rock surface grazing on the green slime (algae). Can you notice any of the limpet's fleshy body? You might see the tentacles sticking out or even part of the mouth if you look closely. What land animal is the limpet like? What happens when you try to pull the limpet off the rock? You can sometimes hear them make a noise when they get a grip on the rock. Can you find a limpet shell with a small hole in it? Is there a living limpet in it?

Probably not. The hole was made by a dog whelk which drilled into the shell with a special tooth, turned the limpet into juice, and sucked up the soup leaving the limpet shell empty.

There's a lot to look at in a rock pool and the surrounding rocks when the tide goes out. Take a good seashore guide with you (see Reading list) and make the most of your time. Ask questions and even if you do not know all the answers you and your child can use your guide together to find them out.

OIL POLLUTION

Every so often a walk along the shore or a swim in the sea will bring you into direct contact with pollution. Oil is everywhere in the form of great lumps of tar just waiting to get stuck to you or your clothes. How does it get there? Why can you never be free of it? It will always be a problem as long as we rely on oil to provide us with fuel to live our present way of life. Try playing the oil pollution game to find out how to stop an oil slick.

OIL POLLUTION GAME

Get a washing up bowl or something similar. A dark colour is more helpful because you can see the oil better. Try pouring some vegetable oil into the water. How much do you need to cover the surface? When oil gets into the sea it creates terrible pollution. People try different ways to stop the oil from spreading or drifting with the tides. You can experiment by simulating the three main methods: the use of barriers or booms to stop the oil from spreading; chemicals to break up the oil or solid material to bind the oil together and sink it so it does not spread.

Experiment with different objects and materials to see which works with your oil slick. String, straws, washing up liquid, washing powder, sand or sawdust can all be tried. If you have some, try out some car engine oil and see if that is any different from vegetable oil. Ask your child what they think might solve the problem.

Which solution would be least harmful to the wildlife in the sea? The next time something like this happens in real life find out what caused it and how the problem was solved. Where does the tar that gets stuck to your feet on the beach comes from?

What's going on?

In reality most of the tar around beaches is oil from ships and boats that is released in small quantities. It builds up over time and never really disperses, drifting onto beaches to spoil our holiday.

SCIENCE AND PETS

An animal at home is an important part of many children's lives, whether it's a goldfish, guinea pig or golden retriever. They make a very good starting point for finding out about the way other animals apart from us work. For this reason 'pets' is a popular theme with teachers and children in school.

National Curriculum note

> The national curriculum at KS1 states 'children should have opportunities both to observe first hand . . . to find out about a variety of animal and plant life. Over a period of time children should take responsibility for the care of living things, maintaining their welfare by knowing about their needs and understanding the care required.'

PROPER PET CARE

In the process of beginning to understand the pet's needs and the urge to take responsibility, children will naturally ask questions to do with the way pets behave. This is the time to talk about the features they have and relate them to their individual needs. Your child needs time to observe and talk through the problems of looking after a pet. Questions like what will it eat? How much food shall I give it? How often do I need to clean it out or take it for a walk? Where will it sleep? Will it be all right when I leave it? All these questions are gradually answered and the proud owner begins to understand the needs of their new pet. This is a good time to pack in as much science as you can. Regrettably, this initial flush of enthusiasm may die down and you may be the one who becomes its keeper.

You might want to capitalise on the enthusiasm being shown for a new member of the household and bring in creatures to study from the garden such as snails, worms or spiders. When the excitement and interest begins to wane simply return your guest pet to whence it came. When the moment is right at some point in the future bring in another set of guests to intrigue your child. Whatever happens, do not bring young fledgling birds into the house if you find them in your garden. The chances are the little bird is being fed by its parents still. It is very difficult to rear young birds successfully. If you do rear it what will you do with it? It will have problems coping in the wild and if you decide to keep it you may need to apply for a special licence to keep the adult bird.

Starting points

In this section we look at things to consider when keeping a pet, questions to ask about pets and bringing in temporary 'pets' to study. We also observe teeth and feeding in pets and how this relates to our own teeth.

Have you ever noticed?

- ▶ Some pets are only active at night.
- ▶ The eyes of cats and dogs point forward; those of a rabbit are at the sides.
- ▶ Mice, hamsters and gerbils chew their cages with their long front teeth.
- ▶ How dogs behave when they meet other dogs.
- ▶ What happens to some animal's eyes when a light shines near or in their eyes in the dark.

LOOKING AT PETS

Make the most of any pets by considering their needs and how these are fulfilled. Discuss with your child what it is you do to look after the animal. What food does it like to eat? How often does it need fresh water? How frequently does it need cleaning out? When does it need exercise? What other things will it need to enjoy life as well as just surviving?

Once you have sorted out what a pet needs you can begin to explore the similarities and differences between different pets. A chart comparing a range of familiar animals and their features as shown below helps to give an understanding of the different animals, their different body forms and how they might live.

The chart will also allow the child to observe quite closely. Where are the ears on a bird – can you see them? Does a goldfish use its nostrils for breathing like us? How do parts of our bodies compare with those of the animals? Why do animals have the kinds of bodies and features they have?

Description	body covering	whiskers	eyes	ears	nose	mouth	limbs	tail
Cat								
Budgie								
Frog								
Goldfish								
Snake	Dry and scaly	None	On side of head	Can't see any	Two nostrils	Wide jaw when feeding	None	Yes

Making connections

Pets are a good way to gain an insight into the world of nature. Consult books, TV programmes and other sources to find out about the way that other animals are suited to their particular way of life.

In natural conditions, wild dogs have little need to have their teeth seen to. They do not eat sugar in such a refined form as pet dogs (or ourselves for that matter) do when they eat choc drops. But teeth maintenance is something we have to keep an eye on because of our sugar-rich diet.

TEETH AND FEEDING

Have you ever noticed how a dog or a cat eats its food? Is it anything like the way you eat? Would you encourage your child to have the same table manners as your dog? Probably not. It's not your dog's fault, more to do with the kind of mouth and teeth a dog has. Bolting large chunks down is the order of the day and chewing bones with the side of the mouth is the only way a dog can deal with food it cannot bolt. Next time you feed a dog watch how it copes with its food. Give it a really big bone to chew on for some really satisfying scientific observations. After it has eaten (it's generally ill-advised to do this before), take a good look at its teeth. How are they different from our own? What are the teeth like which are doing the bone chewing? What damage is done to the bone as a result of the chewing?

WHY DO I HAVE TO CLEAN MY TEETH?

How many times do you get this question thrown at you? Sometimes teeth cleaning can become a nightmare and when you get this question you may wish you could show your child something which will convince them they really do need to clean their teeth without questioning it. Try this and see what reaction you get. Place some pieces of broken eggshell (to represent tooth enamel) into two cups. Pour water over the eggshell in one and vinegar over the other. Leave the eggshells until all the water and the vinegar evaporate. What is left?

You might also like to try this experiment out using other liquids such as fizzy drinks. Are all the drinks the same? What happens with a sugar solution or squeezed orange juice?

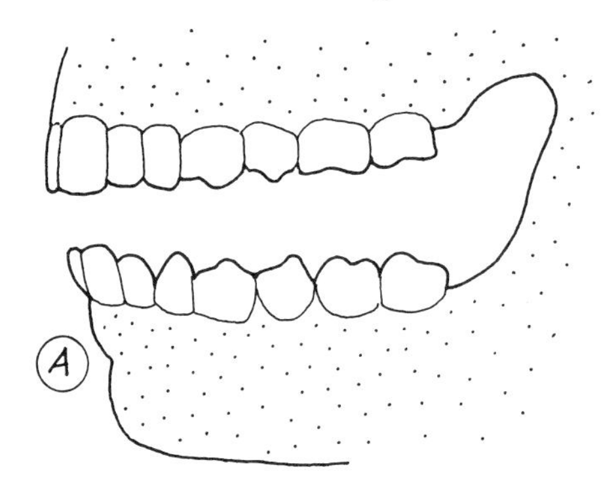

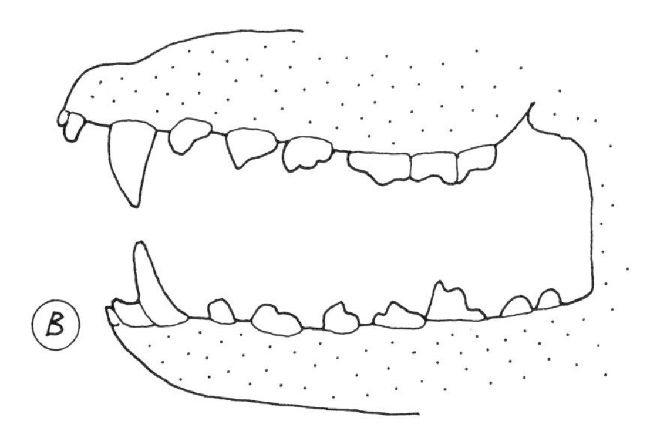

What's going on?

Why this is worth doing is to show the effect of vinegar, a weak acid, on calcium in the egg. Our teeth are made mainly from calcium. The bacteria which breed in our mouths make acid and it is this which wears away our teeth. You can show where the bacteria in the mouth are by swilling some dilute food colouring around in your mouth. Two teaspoons of colouring to a cup of water should do it. The colour sticks to the areas of plaque in your mouth. Try before you brush and after you brush. Will your child now brush their teeth without questioning your wisdom? Will you keep feeding your dog choc drops? Start saving up for the vet's bill if you do!

TEMPORARY GUESTS

If you have a garden or if you have been on a visit to the country you may want to bring into your home some animals to study.

Things to explore

- ▶ Find out about each animal in more detail – what it looks like, how it moves, what it feels like to touch, how similar or different it is to other animals.
- ▶ Discover how it lives – what it feeds on, if it is active in the day or night, how it behaves with others of its kind, whether it prefers damp or dry conditions.
- ▶ Watch an animal and note how it changes, grows and develops.
- ▶ Find out what it needs to live, how to care for it. Plan its needs, be responsible for it as a living thing.
- ▶ Know when to clean it out, find food for it. Change its conditions as new information comes to light. Ensure the animal is returned to its wild home.

Some animals worth collecting include: slugs and snails; worms; ants; woodlice; butterfly eggs; pondlife such as tadpoles.

Keeping animals such as the above is particularly fruitful if you can encourage your child to ask some questions about the animal and then find ways to test their ideas. Snails move slowly, but how slowly? What part of their bodies do they use to move? What do you notice the snail making as it moves along? Why does it need slime like this?

Other animals such as centipedes and millepedes, beetles and spiders can be more difficult to maintain. These are worth a look at to study their main features, but it is probably wiser to return them to the wild as soon as you are both satisfied you have found out all you can.

Basic questions such as 'where is its mouth, has it got any eyes, how does it move and how long is it?' enable your child to appreciate the variety of ways animals can solve the same problems of living.

Observations which your child has made of the way the animal behaves in captivity can help to stimulate ideas for activities and experiments with the animal. For example, an animal may have been staying in a dark corner of the tank during the day. Questions asked around this observation may lead to the need to test the idea that the animal prefers one kind of stimulus to another. One way to do this would be to construct a simple choice chamber experiment.

What you need

- Get a plastic tray and cover half with damp soil, the other half with dry soil. Put a woodlouse in the centre. Which kind of soil does it choose? Keep repeating the test to see whether it chooses the same one each time. Which one does it choose most often?
- Use the same tray to try out different conditions (parameters) such as light and dark, food and other differences. ▷ It is important that you vary only one thing at a time to make the test fair. ◁

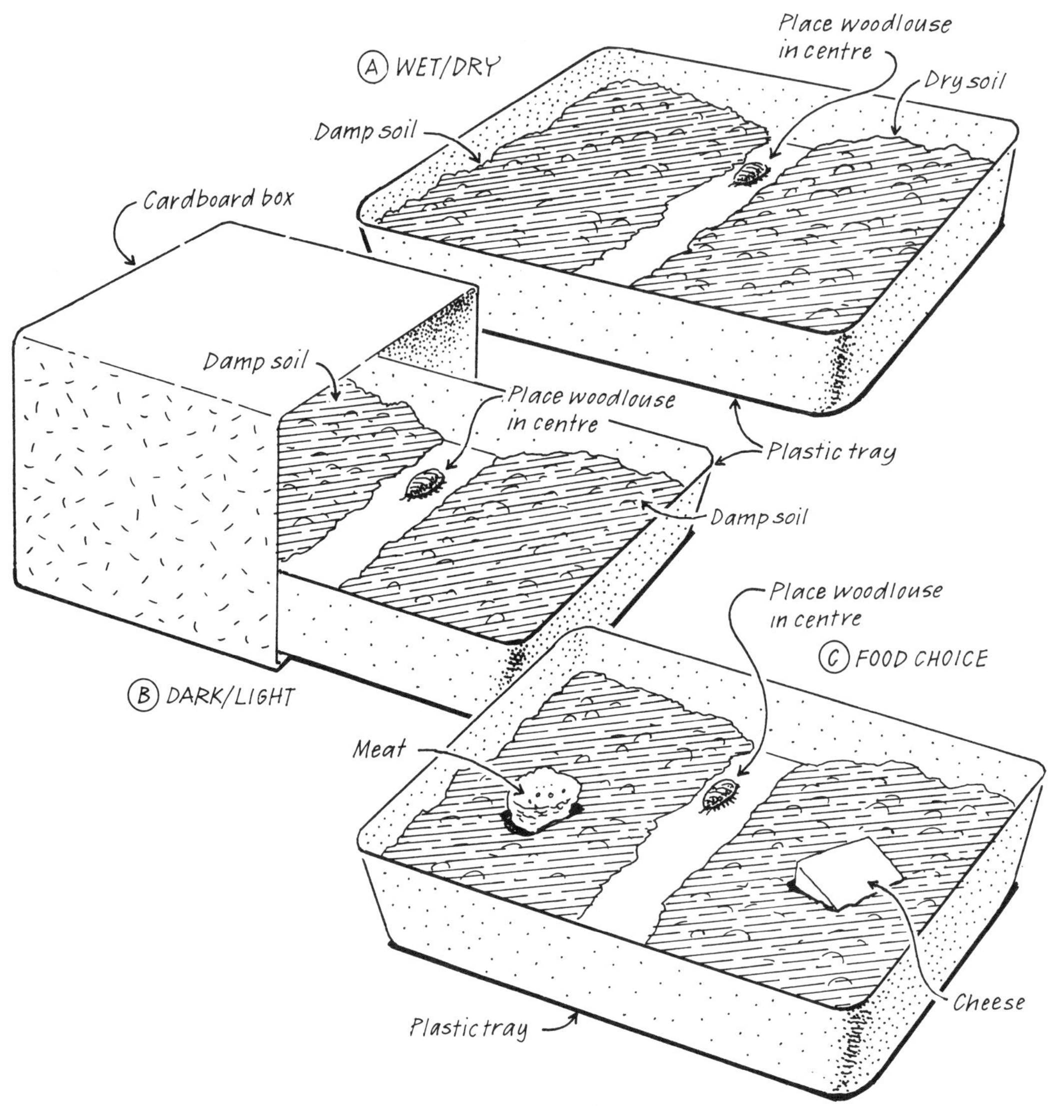

WATCHING GROWTH

In the early spring around March or April frogs lay their spawn in ponds around the country. Collect some spawn, put it into a suitable container such as a small plastic fishtank with some water weed and keep a careful watch on the changes which take place. Look every day to note what is happening and keep a record of the changes. You may have to take a tadpole out to get a good look at how its body changes. Use a magnifier to help get a closer look.

A USEFUL TEMPORARY HOME FOR MINIBEASTS

A useful container for various small animals can be constructed easily from two sheets of perspex or thick glass, some lengths of rubber tubing, plasticine and four to six large bulldog clips. You can put worms or even ants in it temporarily. You can even fill it with water to get a good look at pondlife. Do not keep the animals in the container for more than a couple of hours at most.

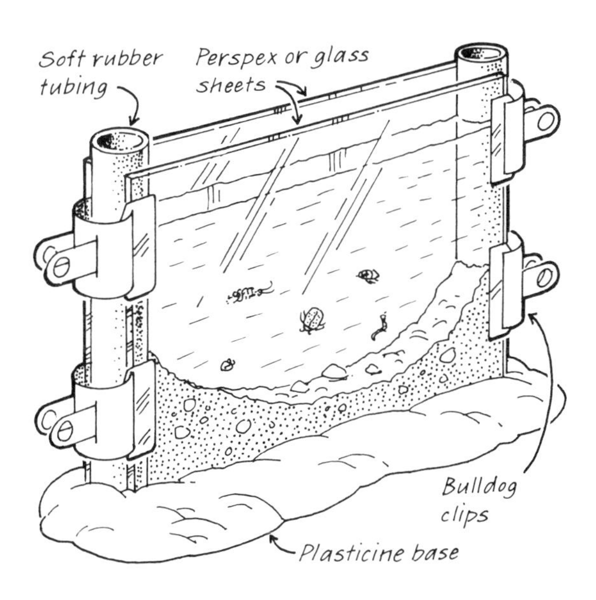

ANIMAL ADAPTATION

Pets also give you an ideal chance to explore a concept which is quite basic to biology but which a lot of children seem to have a problem with, adaptation. There is no need to actually use the word itself and there is certainly no point in doing so with younger children. But it is helpful to think in terms of parts of the animal's body being there because they do a particular job for the animal. We say the dogs feet help it to run quite fast. Not as fast as a cheetah but faster than us. To help it, a dog has claws on its feet and these are out all the time giving it a good grip on the ground – they act rather like the spikes on running shoes. This is an adaptation for running and helps the dog to live its life. But do cats have the same kind of claws? If not, why not? Do they need to run as fast as a dog? Where can a cat go that a dog cannot? Will a dog wear its claws out or do they keep growing? Why does a cat *need* claws it can tuck away except when it wants to sharpen them on your furniture? And where else apart from your furniture does a cat sharpen its claws?

Science and ourselves

For many children, the first living thing they find out about is themselves. The topic called 'Ourselves' is an area of study much used by teachers at school. After all, who else is more important to a child than itself? What better way can a child think about how other living things move and live than by comparing the way his or her own hands and legs operate? So by using activities which focus a child's attention on themselves you can help give them considerable scope for developing ideas through direct experience.

Starting points

In this section we look at similarities and differences between ourselves and others, at aspects of health, the range of our senses, and we also touch on growth and change.

Have you ever noticed?

- How different we all are even though we come from the same family?
- That when we are ill we sometimes get a high temperature?
- When you have a cold, eating is not much fun because you cannot taste anything.
- If you push your ears out from the side of your head it changes how you hear a sound.
- How easy it is to take our senses for granted.

GROWING AND CHANGING

For young children, the notion that growth and change happen can be quite a difficult concept. As children grow they will notice differences between themselves and younger and older children. But it is difficult for them to accept that you as a parent were once a child. As they get older they will begin to notice that they are growing out of clothes or they notice that they can reach up to things they had never reached before without having to stretch. Height charts and comparisons with other older or younger children all help to give them appropriate experiences.

Something to do

Here is an activity which might help. Cut out pictures of people from old magazines. Choose a selection of people which represent a gradation of ages from a baby to an old person. Help your child arrange them into the right sequence.

Things to explore

It's not just growth itself but awareness of their own bodies that very young children will be busy discovering for themselves. Here are some of the main ideas that children can develop during the pre-school years.

- Know what the various parts of the body are called.
- Recognise different body functions both in themselves and other living things.
- Know that there are both similarities and differences between people.
- Come to terms with the limits of the human body by understanding that most human beings can run or jump but they cannot fly; empathise with people who have physical disabilities.
- Be able to understand that the body changes and growth starts with birth and ends with death.
- Understand the importance of sleeping and waking and how they relate to the above.
- Be aware of the different senses through undertaking activities and using language.
- Understand that the body can become sick and tired and in pain.

National curriculum note

The national curriculum states that 'Children should be finding out about themselves, developing their ideas about how they grow, feed, move and use their senses and about the stages of human development.'

ALL ABOUT ME

On the next page is a way of making a chart which will keep a record of measurements about your child. You would probably keep these or similar details in a different form for yourself. Why not keep them altogether. Your child might want to compare them with other members of the family, or with friends. Keeping these records as a measure of how children change is worth doing.
A similar chart could be produced on a large piece of paper, say a piece of old wallpaper, which your child lies on so that you can draw around his or her shape. Cut the shape out and use it to think of interesting ways of representing the information overleaf.

MESSY RECORDS

A fun way to make some of these records is by using paints. Again, a long strip of wallpaper is handy. Using a sponge, dab the paint onto the bottom of your child's feet. They then walk along the wallpaper leaving their footprints. Measure the distance between them for your records.

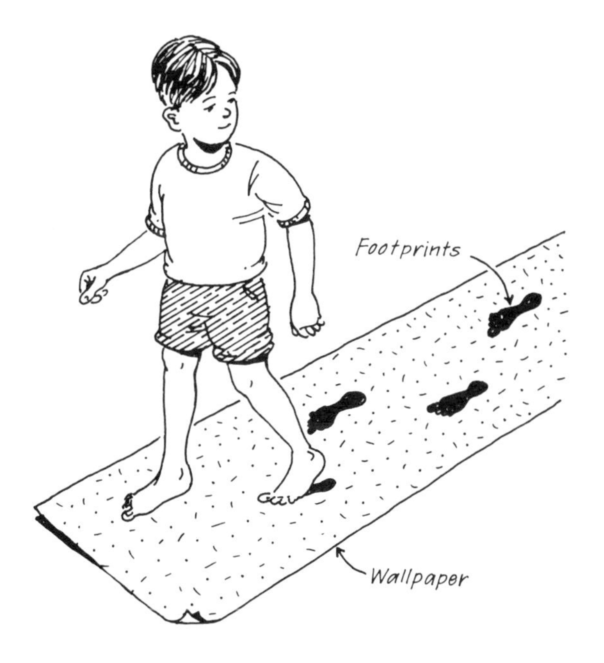

Name: ______________________

Age: ___ years ___ months

Weight: ____ kg.

Height: ____ m. ____ cm.

Circumference of head: ___ cm.

Colour of eyes: __________

Do I have earlobes? ________

Can I roll my tongue? ______

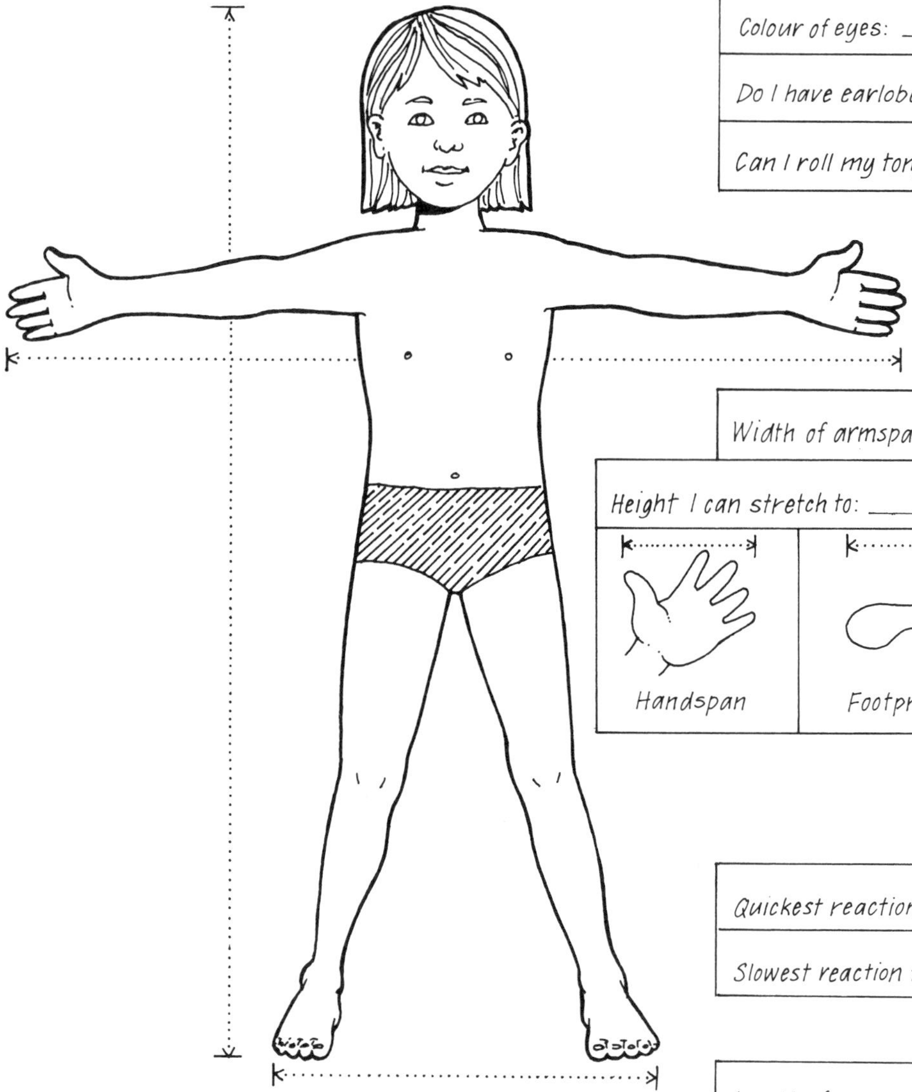

Width of armspan: ___ cm.

Height I can stretch to: ____ m. ____ cm.

Handspan

Footprint length

Quickest reaction time: ______

Slowest reaction time: ______

Length of one pace. __________

FINGERPRINTS

With fingerprints you will need to experiment with the amount of ink. Collect prints from a number of different people. Look for the four types shown below: whorls, loops, arches and composite. Are there any other kinds? Are any of the prints the same on two people? Once you have made the prints take some more from one of the people whose prints you have and ask your child to say whose prints they are by comparing the ones on record. How easy is it to compare?

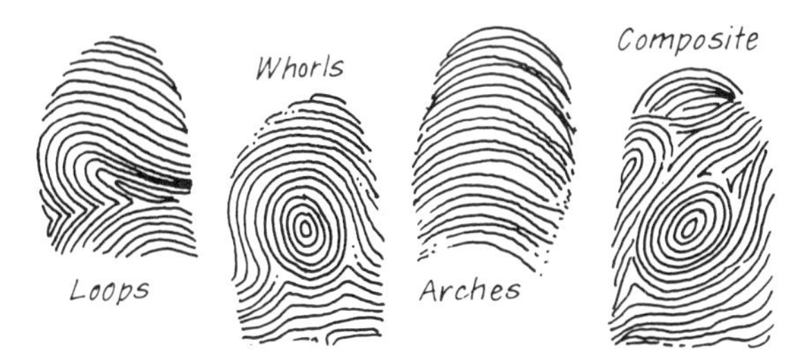

FAMILY LIKENESSES

Features such as the ability to roll your tongue, the presence of a prominent ear lobe, the colour of your eyes and hair, are all inherited features. On family visits the same information could be collected to find out which of the family members share the particular features. Who has ears like your child's? Who can or cannot roll their tongue? Who has the fastest reaction times?

QUICK THINKING

To find out who has the fastest reactions, you can measure them with a ruler. How? You hold a ruler upright. Zero should be at the bottom. Your child holds their hand at

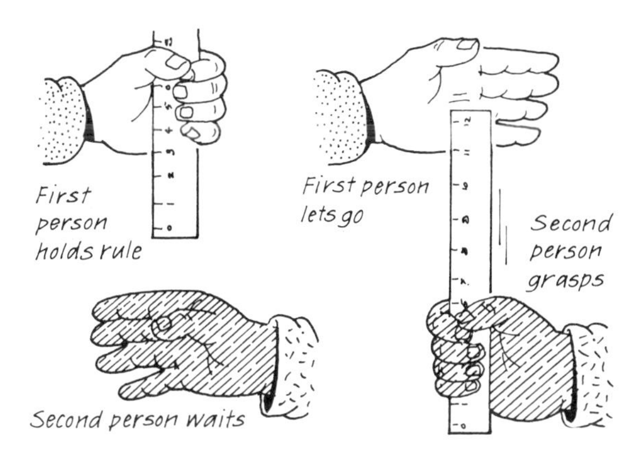

the zero mark, ready to catch the ruler as it falls but their hand must not touch the ruler. You drop the ruler suddenly and the child must catch it. Record the point at which they catch it. Repeat several times. Do they improve at all? You swop around. Are you any quicker? Do you improve at all? Might there be a difference caused by age?

OUR SENSES

We use different senses to come to understand our world. This activity helps you think about the senses we use and helps your child realise the range of those senses. Below is a list of paired objects. Each pair is identical except for one feature. The idea is that your child explores the objects to try and discover the difference. They then have to tell you what they did and which sense or senses they used to work out the difference. You can watch your child carefully to find out how they explored. For example, the two jars with the cotton wool in will need to be picked up and shaken. You could come up with some of your own pairs of objects – here are some suggestions:

- Spot the difference puzzle from a magazine.
- Two similar objects but a different colour.
- Two cards of slightly different brightness of white.
- Two jars with their lids on (these should not be removed by your child) with cotton wool inside. One jar has the cotton wool stuck onto the bottom.
- Two matchboxes, one with a small ballbearing and the other with a piece of cork, the idea is they should feel similar but sound different.
- Two whistles or bells which look the same but sound different.
- Two similar pieces of material which feel different.
- Two pieces of cloth, one with a nice smell one without.
- Two jars or plastic bottles, one with hot water the other with cold.
- Two glasses of water, one with orange juice the other with food colouring. How do they taste?
- Two matchboxes, one filled with lead to make it heavy, the other empty.

The list of objects you can use is endless. The process of coming up with your own suggestions will help you think about the way we all explore the world.

FOOL YOUR BRAIN

When you have explored the different ways in which we make sense of the world you can try the following activity which gives an inkling of the way in which our brain cannot quite so easily cope with the information we are given.

Take a piece of thin card and fold it down the middle. Place the card on a table and look at the folded edge with one eye closed. What do you notice? Is the paper folded in or folded out?

What's going on?

Your brain cannot decide what is happening. That's because the open space and one closed eye makes it difficult to judge depth. Your brain tries to see the card in different ways and keeps changing what you see.
This is an optical illusion and there are lots of them. Several of the books included in the reading list have some of the more well known ones.

EYES AND VISION

We have eyes on the front of our face. That's handy because it helps us judge distance and see depth. Our monkey-like ancestors found that particularly helpful living up in the trees. Cats too find it useful when they pounce on their prey. But what happens when you use only one of your eyes?

Try this.

Put a cup on a table and ask your child to stand about 3 metres (10 feet) away. Ask them to cover one eye with a hand. You stand slightly in front of the cup with your arm out holding a coin above it. Your child should be looking at the cup and the coin. Tell them to give you instructions on which position to move your arm in order to drop the coin in the cup. How good a shot are they? How good are you for that matter? How much practice does it take to get it right? What clues are you using to help you judge distance?

What's going on?

The distance between our eyes means our brain gets a slightly different image from each eye and when comparing those images it gives us a three-dimensional picture. When you cover up one eye you lose the ability to see in two dimensions and you cannot judge distance so well. You can begin to compensate though by taking into account the size of other objects nearby and their brightness.

TOUCH

What can we tell from using just our sense of touch? Cut a hole in the side of a cardboard box and fill it with various objects. Try fruit, a potato, a building brick, a feather, a leaf, a cube of jelly, a pasta shape, the list is endless. Have your child feel one object at a time. Ask them what it feels like and get them to describe it. Is there anything about it which tells them immediately what it might be? Tell your child to predict what they think it is and then have them remove it. Were they correct? How many did they work out? Try it again using some different objects but put a couple of the first set back with the new ones. Did your child remember the feel of these or did they think they were something different?

Which part of our bodies are the most sensitive to touch? Put two pins in a cork a slight distance apart. Now touch the heads of the pins gently against different parts of your child's body. Places such as the flat of the hand, on both sides of the arm, the back of the neck, the fingertips. Ask your child how many pinheads they can feel each time you do this. Which parts of the body are more sensitive than the others?

What's going on?

The skin registers both pinheads where there are many touch receptors (the nerve cells in the skin which register touch) close together. Where there are few touch receptors they feel only one pinhead.

SPOT THE PULSE

Children can detect what is going on in their own bodies by feeling for their pulse, listening to their heartbeat and feeling it beat, particularly after some exertion. Here's a suggestion for seeing a pulse in your body. Push one end of a 10 cm (4 in) piece of drinking straw into a piece of plasticine. Stick a drawing pin into the plasticine-filled end. Rest your child's arm on a table with their hand facing upwards. Place the pin onto the artery on the inside (thumbside) of their wrist and watch what happens. The length of the straw amplifies the movement of the blood flow in the artery, making it very obvious. You will find it easier to count the pulse rate. Where else can you see or feel a pulse rate in your body?

A QUESTION OF GOOD TASTE

Can your sense of taste recognise different flavours of crisps? Get some different flavoured crisps. Place each flavour on a different plate. Make sure you have labelled each flavour but keep the label hidden under the plate. Ask your child to taste the crisps. Can they put a name to the taste in their mouth? If they have problems working them out let them try each kind of crisp and tell them what flavour they are. Now rearrange the crisps on the plates without them seeing and try again. Can they get any more right after having tried this? Have they learnt the tastes?

You could try a similar test with some sweets such as fruit gums. Ask your child which is their favourite flavour. Get three different sweets including their favourite one. Tell them to close their eyes and put the sweets in their mouth one after the other. Give them time to chew each one in turn. Can they recognise their favourite one after all? If they cannot, what other sense is playing a part in helping them choose their favourite flavour?

Try this taste test using fruit and vegetables. Cut up cubes of peeled raw apple, pear, potato, carrot, turnip, and onion. Make sure you cut the pieces up into cubes of approximately the same size. This time blindfold your child so that they cannot see what they are eating. Place a single cube of food at a time on their tongue. Then get them to hold their nose whilst they are eating. Can they tell what they have in their mouth? Now you try it. How many did you get correct between you?

What's going on?

In the case of the crisps, we cannot always give a name to the flavour. With the fruit gum test, the colour and what we expect a colour to taste like from previous experience, all play a part in helping to choose a favourite flavour. Our taste buds, however, are just not that subtle. The taste buds in your mouth can only taste sweet, sour, salty and bitter flavours. The rest of the information about flavour comes from another sense – the sense of smell. After all, what tells you that there is something good cooking in the kitchen? Smell and taste are closely connected. Basically, smell is a long-distance sense and taste is a close-up sense.

SMELL

Try out your sense of smell with smell pots. Collect some yoghurt pots and cut out some circles larger than the opening of the pot from a pair of old dark-coloured tights. Secure the tights over the top of the pot with a rubber band. Place different smelly things inside the pot. Choose a selection: orange, lemon, chocolate, cocoa, coffee, soap, lavender, mint, different herbs and spices, and so on. Which smells are strongest? Which are easiest to identify?

HEARING

Ask your child to push their ears out and listen to a source of sound. Do they notice any difference between having their ears out and flat against their head? Try making a cone-shaped ear trumpet out of card and sticky tape or use a large household funnel. Have your child place this against their ear. What happens? Tell them to move the trumpet around listening for different sounds. They should hear the sounds more clearly.

When your child is looking at books, watching pet animals or visiting the zoo ask them to look at the ears of other animals. Do all animals have ears like ours (ear flaps or auricles as they are more correctly known)? Do the animals that have them always have the same sized ear flaps? Now that they have tried out extra large ears can they work out which animals are likely to have good hearing?

The ear flap or auricle is the part of the ear which collects sounds. The sounds themselves pass down into the ear and hit the eardrum, making it vibrate. The following activity may help your child grasp the idea that air pressure can make something move and vibrate. Construct the apparatus shown below using a round cardboard container. It must have a bottom, and a top with a hole in one end. Place the hole near the flame of a candle. Now have your child flick the other end with their finger. They will find they are creating a sound. The candle flame will move and flicker. By flicking the end of the container very hard they may be able to put the candle out altogether.

What's going on?

The sound your child created in the container is caused by disturbing the air inside, which then disturbs the candle flame. Our own ear drums respond to disturbances in the air (sounds) in a similar way.

GOOD VIBRATIONS

Sound doesn't always travel through air. Try making a telephone from two yoghurt pots connected by 5 metres (about 15 ft) of string. Keep the string taut. Your child listens at one end and you speak into the telephone at the other. Experiment by changing the size of the yoghurt pots, the length of the string and the kind of string you use. Which works best of all?

Can sound travel in water? Next time you go swimming with your child ask them to listen out for sounds under water. What can they hear? If you can't get to a swimming pool fill up a balloon with water. Put it to your child's ear and then place your lips against it. Now make some high and low pitched sounds. Which were easier for them to hear? Now try it with an ordinary balloon filled with air. How different does it sound?

All these activities help your child to begin to appreciate their own hearing ability, the design of ears in other animals, and the nature of sound itself.

HOT HEADED

When children fall ill they may have their temperature taken and wonder why. Sometimes it's just a hand on the forehead to try and gauge temperature. Or it may be necessary to use a thermometer. When the thermometer is being read, terms like high and low are being used. When children are told they have a high temperature they probably find that doesn't fit too well with feeling cold when they are wrapped up. Although this is not the best time to be thinking about science, it can provide some food for thought at a later stage. Discussions about feeling hot, staying warm when it's cold, and body temperature can be begun. Below are some activities which explore this area.

HOW HOT DO I GET?

The thermometer is an excellent scientific measuring instrument to have about the house. You do not always have to reserve it for when someone in the family is ill. Take your child's temperature. Show them the scale and help them read it. When your child has not got a cold it will be around 37 degrees C (98 degrees F). This figure is actually based on average temperatures so it may vary slightly from person to person.

Ask your child if their temperature will rise if they dress up in warm clothes and then run around. Get them to do this and when they feel hot take their temperature. It will not be very different. Apart from hot, what else are they feeling? They may be sweating and this is the way we stop our temperature from going high. They may recall sweating when they were ill in bed sometimes without having to run around. This can help you explain that we are warm blooded creatures like all other mammals and birds. All other animals are cold blooded.

Something to do

When you placed your hand on your child's forehead using your hand you were trying to sense their high body heat. Here is an experiment which shows you why that way can be misleading. Get three bowls large enough to put your hands in. Fill one with cold water, the other with warm, and the third with hot. Place one hand into the cold, the other into the hot. Leave them there for a couple of minutes. Now put both hands in the bowl with the warm water. What can you feel? The water temperature in that bowl is the same even though one hand tells you it's cold and the other hot. The sense receptors in your hands get used to the temperatures of the first bowls of water. They are fooled into giving your brain misleading information when you place your hands in the warm water bowl.

KEEPING WARM

What is needed to keep something warm? Fill two jam jars of the same size with hot water. Then ask your child to wrap one of the jars up with materials they think will keep the heat in. Use the idea of clothes if they are not sure what could be used. A thick scarf, a woolly hat or some gloves can be wrapped around one of the jam jars. You need to wrap the top and bottom of the jar as well as the sides. Leave the other jar free of wrapping and wait until it cools right down. Now unwrap the other one and feel the water. Has it worked? If you have a thermometer you could take the temperature of the water in the jam jars before and after the test.

There is another dimension to this experiment. Would wrapping up a jam jar full of ice melt it quicker than without? Try it. Take two jam jars and fill them both up with ice. Wrap one completely and leave the other. Keep them on a table top or, on a sunny day, on a window ledge. When the ice in the unwrapped jam jar has half melted, unwrap the other. What has happened? The ice will have hardly melted. The wrapping has acted as an insulator stopping the warmth of the room or the sun from melting the ice.

National curriculum note

By KS 1 children should explore the effect of heating common everyday substances, for example, ice, water, wax and chocolate, in order to come to an understanding of the role of heating and cooling in bringing about melting and solidifying. They should begin to link the feeling of hot and cold in, for example, water, their own bodies, air, with temperature measured by a thermometer.

By KS 2, children should be able to investigate changes that occur when familiar substances are heated and cooled as well as hot and cold in relation to their body temperature.

KEEPING CLEAN

Much of what children learn about basic hygiene, they learn at home. Washing hands before handling food, for example, making sure the fridge is kept cold and throwing away mouldy food. They will also be interested in your procedures for discouraging the growth and spread of bacteria and disease: covering up food to stop flies and bacteria landing on it; using a different chopping board for cutting up raw meat from the one we use for cooked meat or cutting the bread. Tell your child why you do these things and encourage their inquisitiveness.

A VISIT TO THE DOCTOR'S

As they get to know their bodies, we should encourage children to be confident about telling others when they notice something is not quite right about them. We have been scientific when it comes to exploring the finer workings of the human body so there is no reason why we cannot be the same when it comes to letting the doctor know about an illness. Yet sometimes people are afraid to tell the truth about their sickness to a doctor. Doctors are people like anyone else and therefore need accurate information before they can apply their scientific training and make a diagnosis. Both you and your child can communicate scientifically on these occasions by giving accurate information about symptoms. Where is the pain? How often does it occur? What kind of pain? Are there any signs of a rash on parts of the body the doctor cannot see? All this helps in the process of diagnosis. When you do not understand what a doctor is saying ask for explanations back in language which you can understand.

If you need more information to help you understand diseases there are often leaflets in the doctor's surgery which give helpful information and which you can take away and read at your leisure.

Making connections

Being ill can lead to thinking about what makes us unwell in the first place. Some diseases can be transmitted through coughing and sneezing, just one route by which some bacteria pass from one person to another. Simple actions such as placing the hands over the face when sneezing give us the opportunity to discuss how science is part of our everyday lives.

SCIENCE, TOYS AND GAMES

You'd be surprised at how well science activities go down at children's parties. Some of the machines, constructions and games described here make good toys for any time as well as providing an intriguing distraction for idle moments; some of them could end up absorbing the family's attention for a whole afternoon. As with all activities with children and young people be very careful to judge the mood and the moment. If it's one of those afternoons when everybody is as high as kites then you would be better off having a balloon fight than trying to do anything too subtle. No one is better at judging these moments than parents themselves.

Starting points

In this section we give just a few examples of the many things you and your child could make and do together. Many more examples can be found in children's activity books (see Reading list). Children gain a lot of confidence in being able to perform some of these activities themselves as tricks and in improving the ideas we have suggested here.

DEEP SEA RESCUE

This is a good one to perform at children's parties if you can get everyone's attention. Send your own friendly diver down to the bottom of the bottle and get him to come up again at your command (see illustration overleaf). If you are really expert you can rescue a drowning paper-clip.

Take the top from a marker pen – a child's felt tip will do well. Cut out the the shape of your diver or, if you prefer, a fish shape, from the lid of a margarine tub. Colour it in for effect. When you are satisfied with the colouring, fasten it to the pen top with the kind of glue that sticks plastic. These types of glue by the way are usually best kept away from little fingers. While that's drying prepare a polythene drinks bottle by stripping off the label so that you get a good view inside all the way round. Fill the bottle to the brim with tap water and have ready a measuring jug and some modelling clay.

Try and float the diver in the measuring jug as in the illustration. The pen top should float if you manage to get some air trapped in it. Stick some modelling clay on the outside to help keep your diver upright. You can adjust how high in the water it floats by either trapping more air or using less modelling clay. If the pen top has got any extra holes in it you can plug these with plasticine, but try and plug these before it gets wet otherwise it's difficult to get a good seal.

Now put your finger over the open end of the pen top so that no water escapes and keeping it there lift the pen top out of the measuring jug and put it into the full bottle.

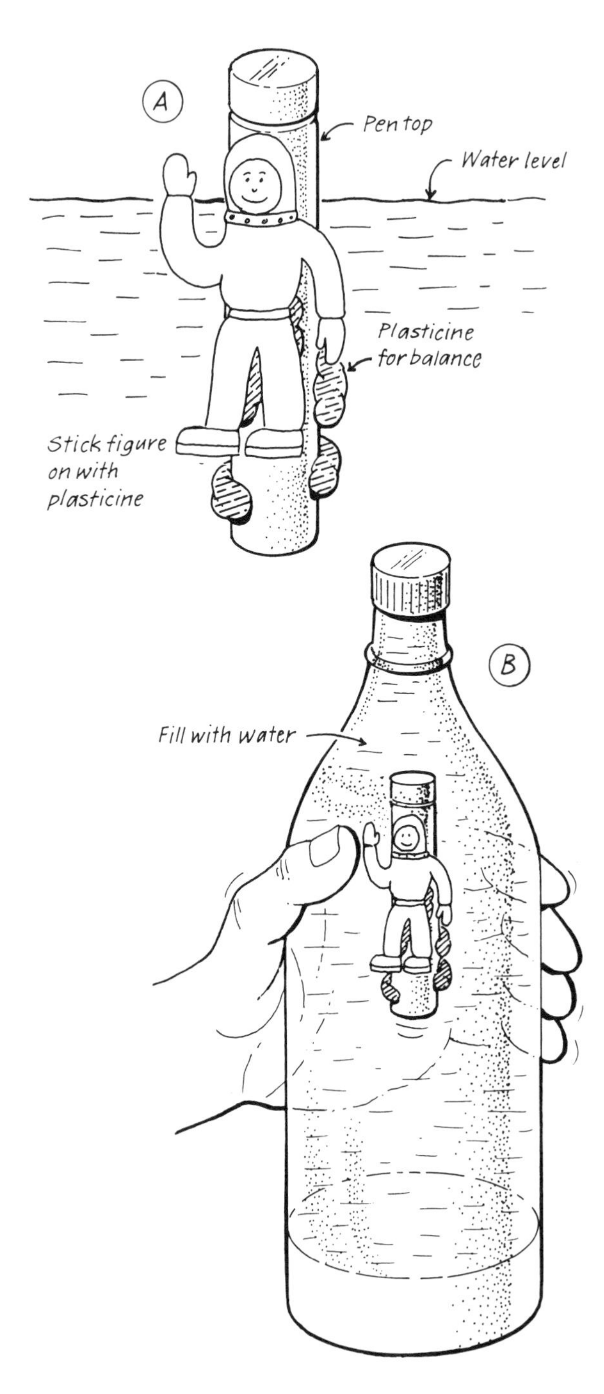

You might have to fiddle about getting the diver's legs in. Once the open end of the pen top is beneath the surface of the water again you can remove your finger.

Some water is bound to have spilt out of the bottle while you were doing this so top it up again from the measuring jug and screw the bottle top on tightly. It is important that there is no air trapped in the bottle except that which is keeping the diver afloat.

Now squeeze the sides of the bottle. As you squeeze the diver will descend. When you release then the diver comes up again. You can impress a young audience by holding the bottle and speaking to the diver as you squeeze. If you hold the bottle in two hands then you can make your squeezing almost imperceptible if that's what you want to achieve. Let the children try for themselves and give any help you feel you want to to maintain the enjoyment. Don't let anybody knock the bottle over, usually the bubble of air escapes from the pen top and the diver lies helpless at the bottom of the bottle.

You can make a more sophisticated version of this toy by sticking a magnetised paper clip behind the diver's boots. You can then use this to rescue small objects from the bottom of the bottle.

What's going on?

Water doesn't squash very well, it's what is known as incompressible. When you squeeze the plastic bottle with your hands the only thing in the bottle that can give way is the air trapped inside the pen top. This means that the air is going to take up less space in the water which means it's going to sink.

BUBBLE MACHINE

Take a loop of wire. Have ready a fairly strong solution of water and washing up liquid. Cut a fine piece of thread and tie it in a loop – this must be smaller than the wire loop. Trim the ends of the knot and try and keep the whole thing clean.

Take the wire loop and dip it in your soap solution and get a bubble to form across it. Then gently place the cotton loop on top of the soap film as in the diagram. Try not to sneeze while you're doing it.

Now, try and predict what might happen if you just burst the bubble inside the cotton loop. Do it. Let everybody have a go.

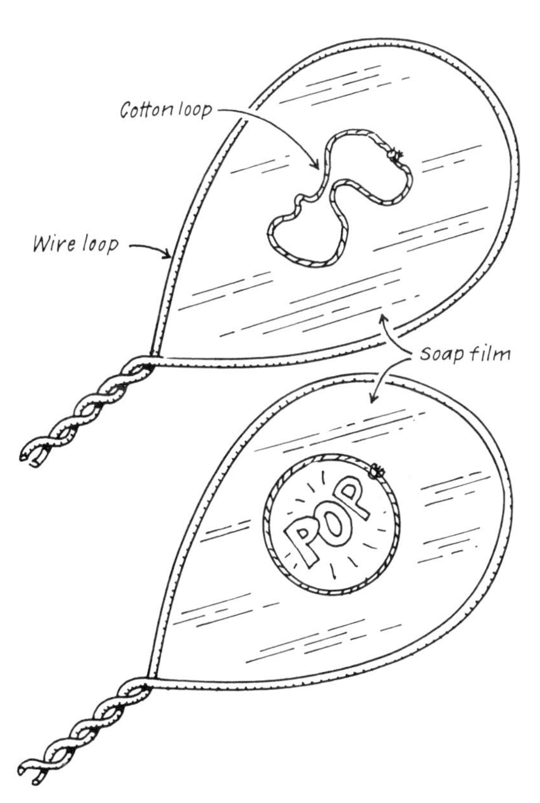

What's going on?

Molecules are attracted towards each other. When the cotton is lying on the soap bubble then there are molecules of bubble all over the place so nothing much happens. If you break the grip of the molecules on the centre then the cotton loop is pulled out all the way round by all the other molecules.

COINS IN A GLASS

Take a glass tumbler of water and fill it to the brim. Beside it place a pile of coins. With practice you'll know just how many. Ask your audience if they know how to put all the coins in the glass without losing any water. Once you've agreed it's impossible then gently slide them in one at a time and surprise everyone.

What's going on?

Water molecules are attracted to one another. This is used to explain the shape of water droplets you see dripping from the tap or raindrops on glass. If you look at the surface of the glass of water after you have slipped the coins in you can see that it is bulging. As you put in more coins then the more it bulges until the force of attraction between the molecules cannot hold onto them any more and the molecules fall over the side. This force of attraction is smaller if the surface is dirty so make sure that the glass is clean and if you want to be successful, keep your greasy fingers away from the rim.

MERRY-GO-ROUND

Using the engine you made in the *Town* section (page 53), you can convert it into a merry-go-round that can provide a good centre-piece for the table at a children's party. You can borrow bits from any construction kits you have to make even more elaborate versions.

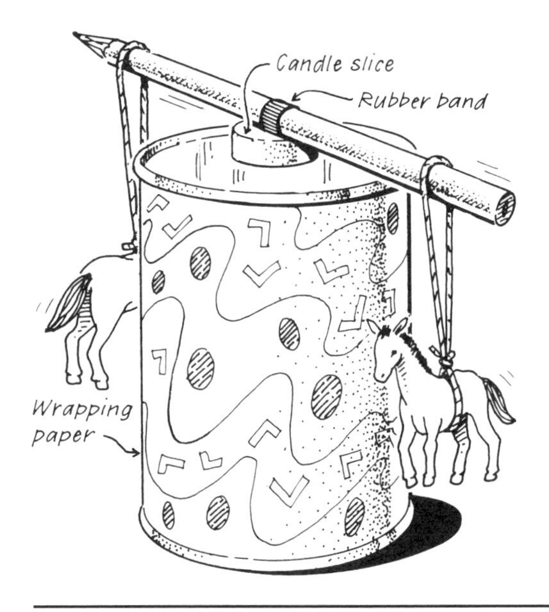

LOOKING THROUGH WATER

'Hey look at my straw! When I put it here it's normal and when I move it over the other side it goes all big'. This conversation was overheard from a six-year-old girl who on another occasion had us both completely absorbed gazing through a bowl of water for almost two hours. One thing that's never in doubt is the fascination children have for water. Whether it's playing in the bath, just simply watching the waves on the beach, or in this case looking not at it but through it.

You will need to choose the moment for what follows wisely since children will need patience in order to look carefully. A party is therefore not the ideal setting for this particular activity. Neither will it be so rewarding if children are tired. However, it is an excellent activity for keeping a group of two or three children of six and upwards occupied for some time. Below this age children may have some difficulty relating to what they see. You will probably find that at this age they will constantly seek your attention too!

Things to explore – 1

You will need a clean milk bottle, a piece of white paper or card, and a nice dark-coloured marker. Draw an arrow on the card the same size as the one shown here. Fill the milk bottle with water and look through the bottle at the arrow. Now slowly move the arrow away as you keep watching. What changes do you notice?

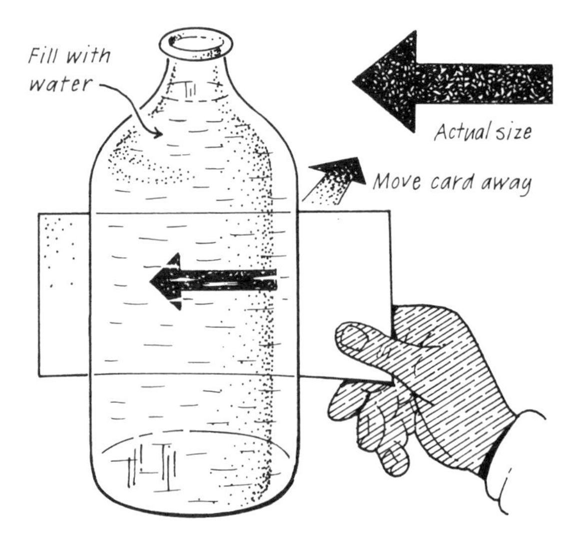

Try it with an empty bottle. What about using different jars? How does the arrow behave with a narrow pot compared to a wide jar? Do you get the same effect when you look down through the jar instead of through the side?

Things to explore – 2

Take a pyrex mixing bowl and fill it with water. Look through the side and drop in a coin – where does it disappear to? Where do you have to put your head to see it again?

Draw this shape on a small piece of white paper and move it up and down as in the diagram to see a fierce mouthful of teeth opening and closing. Let children draw their own pictures of this effect.

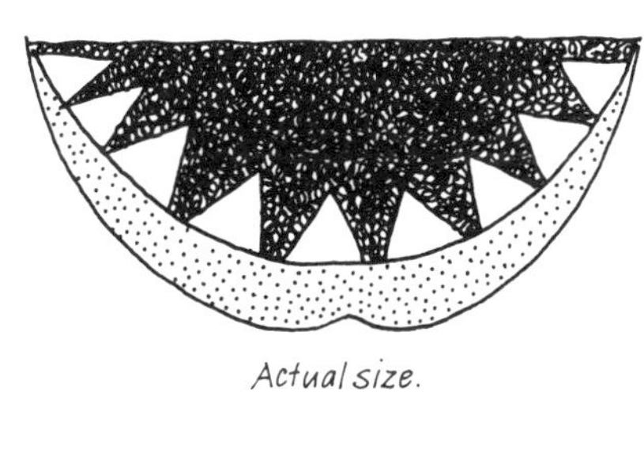

Actual size.

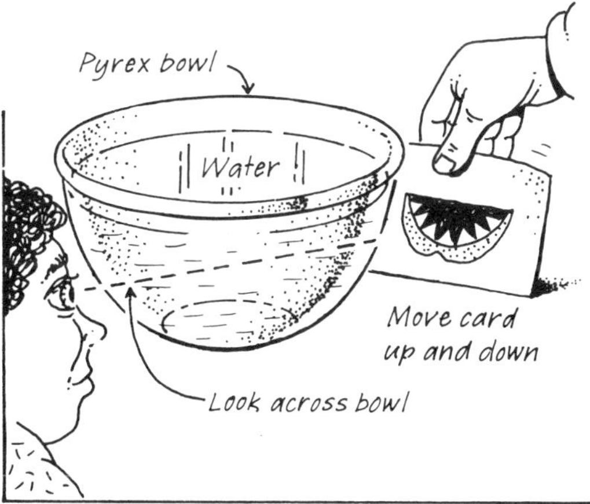

Look up at the surface of the water from underneath. It seems to act like a mirror; in fact you can use it to do mirror drawings as shown here.

Try poking a pencil or finger in the water while you are looking up; from what you see, work out how to make this concertina play (silently of course).

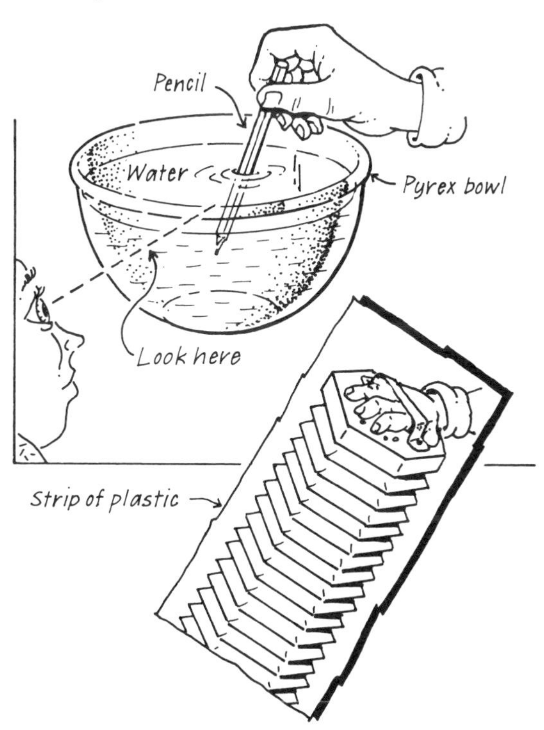

Can you see any colours? Try looking through with the light behind you or look at the paper on the other side. Try looking up through the edge of the water. Try making ripples as you do it.

Catch rainbow colours on white paper

Safety first

Avoid looking at strong light through glass and never look directly at the sun.

A TIN THAT MOVES

Young children like rolling things about; older children will do it at the meal table anyway. Try wedging a wodge of plasticine on to one side of the inside of a cocoa tin and watch it dance about with a mind of its own. Alter the amount of plasticine and the way it's distributed.

ROCKET LAUNCHER

If you know you've got a party on the way then it's worth saving up a few odds and ends. A few empty (and well rinsed out) washing up liquid bottles will enable you to hold a space launch competition (see illustration below), ideal entertainment for a group of six-year-olds.

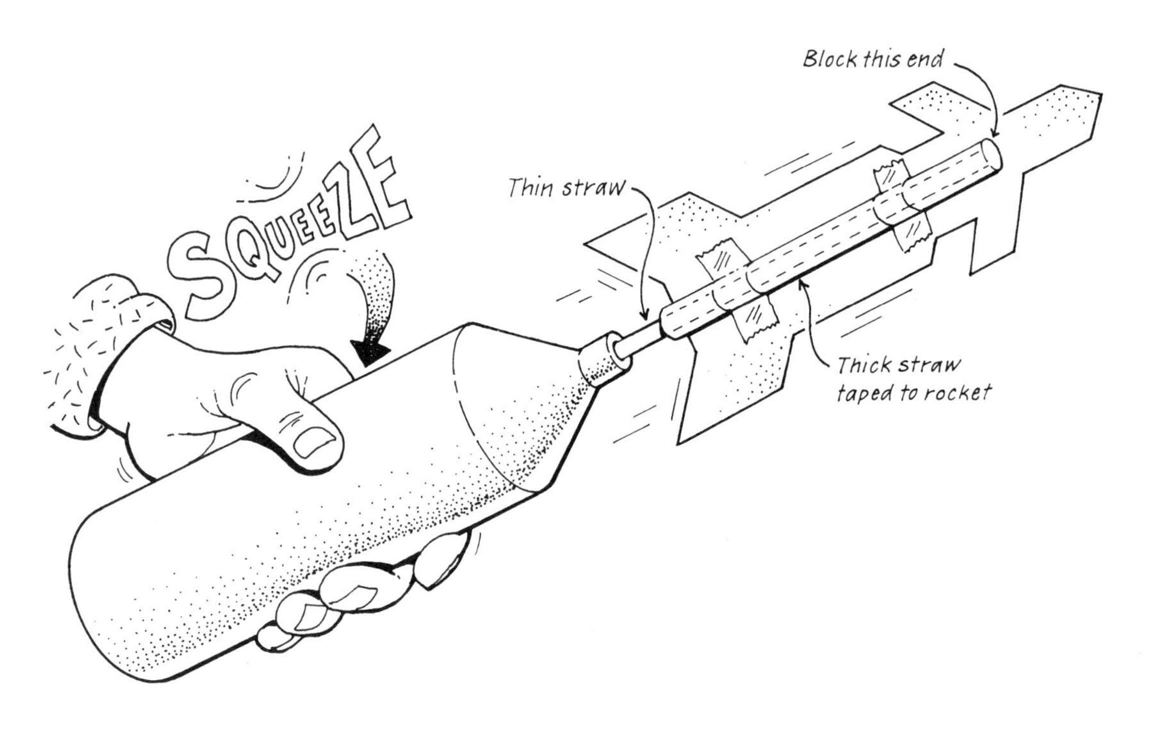

A MUSHROOM THAT SPINS

Try making these spinning mushrooms. You will need a cut-off ping pong ball (or the top from a foam bath bottle), a pen top and plasticine. Adjust the distribution of the plasticine until you can get the mushrooms to flip onto their stalks as they spin of their own accord.

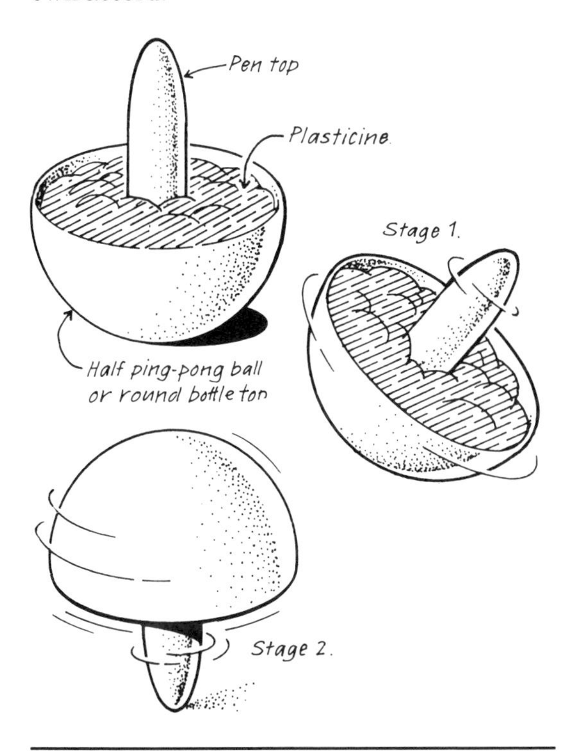

SCIENTIFIC MUSIC MAKING

A simple experiment to do at home and introduce the link between science and music. Get hold of a number of milk bottles. Fill them to different levels of water (see front cover). Now you can either blow across them or tap them with a spoon. Tune your bottles by adding different amounts of water.

INSTANT MUSICAL INSTRUMENT

An old tissue box with an oval hole in the top where you take out the tissues makes a good 'musical box'. Place rubber bands of various thickness over the length of the box. Now twang them to create a guitar or double bass sound.

All the sounds are caused by air moving. In the bottles, varying differences between the tops of the bottles and the water make different sounds. In the tissue box the sounds are created by the rubber bands vibrating the air inside the box.

Making connections

Many of the activities in this book can be modified to be the focus of a new game or toy. If you visit a hands-on science centre or browse through activity books (see Reading list) then you will find many more ideas. If toys and science interest you then some polytechnics run informal courses that are specially designed for you (see Useful organisations for more details).

Going on from here . . .

Now that you have read this book and tried out some of its suggestions you may want to pursue things further by following up some more of the suggestions in the Resources section. Or you may simply want to continue to potter about in your own kitchen sink. Whatever you choose to do we hope that this book has helped you and the child or children in your care to feel excited by the world of science and more confident in continuing to explore it.

SCIENCE AND THE NATIONAL CURRICULUM

WHAT IS THE NATIONAL CURRICULUM FOR ENGLAND AND WALES?

The new national curriculum came into effect in August 1989 as a result of the Education Reform Act. Put simply, as soon as children attend school, or even before, the national curriculum acts as an education map, with signposts for teachers and parents to help them be aware of how a child is progressing, how their ideas are developing, and how best to support and encourage them further. Since children spend many years of their lives at school and they all have individual needs, the national curriculum acts as a convenient framework to help with the planning of ideas and skills both in the classroom and – since the Education Reform Act calls for greater collaboration between parents and schools – out of it, in the home. The aim of this framework is to build a progression of skills and ideas which will develop from the ages of five to sixteen – and hopefully beyond.

If fully and successfully implemented, the national curriculum for science should ensure the following:

- that science is available to all;
- that there is balance and breadth in the science children learn;
- that it appeals to a range of aptitudes and cultural backgrounds;
- that it equips children for life as well as for jobs.

HOW, NATIONALLY, DOES IT OPERATE?

Each national region of the UK has its own administering body for education and each therefore has its own arrangements for implementing the curriculum. Only the detail is different. For example, in Scotland, science has been an entitlement for children of primary school age for some time, but unlike the other regions, it is not organised as a separate subject but forms part of the curriculum for environmental studies within the new 5 to 14 development programme in Scotland. Despite these differences, common to all are the sorts of ideas about science and the approach to doing and learning science which children will encounter from now on.

THE EDUCATION REFORM ACT AND PARENTS

The Act specifically makes provision for parents to become more closely involved with their children's schooling. You may already be doing this, going into school to help with reading or maths, or as a member of the school parent-teacher association. You might now think of other ways of becoming part of the wider school community.

It may be possible, for example, to get children involved in the logistics and technology of an event such as a summer or Christmas Fair. They could help plan the layout, the communication system, or the signs that will direct visitors to the displays and games.

WHAT (OR WHO) IS THE NATIONAL CURRICULUM COUNCIL?

To comply with the Education Reform Act the national curriculum council acts on proposals from the Secretary of State for Education, and keeps the curriculum frameworks for the various subjects schools cover under review. These are then presented before Parliament and, if approved, they become law, published as Statutory Orders (SOs). Information about the national Curriculum is also available as a separate publication, *Science in the national curriculum*, published by HMSO, price £3.80 (£6.95 with ring binder). Three subjects (the so-called core subjects) now have their own SOs: science, maths and English (Welsh in Wales; Northern Irish in Irish-speaking schools). These represent your child's entitlement by law to education in science. It is important to remember that they are not a set of hoops for children to jump through. The onus is on the education system – usually the school governors – not on the children themselves since learning is (and remains) the prerogative of the learner, not the teacher.

The principles written in to the national curriculum are applied nationally, although schools themselves are able to determine how subjects are taught. The three core subjects, English, maths and science, will now take up considerably more of a child's time in school than the other subjects. Science therefore forms a central part of the minimum education requirement since it is one of the core subjects. Other subjects (also the subject of forthcoming legislation) are known as foundation subjects. The national curriculum for science sets out the sorts of understanding and experiences children should have the opportunity to develop in school. These are defined in the form of *attainment targets* (ATs) and *programmes of study* (PS) and cover the education of all children in the age range 5–16.

HOW DO ATTAINMENT TARGETS WORK?

Attainment targets provide teachers with a good way of establishing what stage a particular child may be at in relation to a particular set of skills or ideas. They also provide a means of communicating this to you, the parent. Of course the view you have of your own child also counts – and this does not simply depend on ATs. This issue is one where there will be continuing discussion between home and school.

In operation, the knowledge, skills and understanding that children achieve at any particular stage are assessed on a ten-level scale from ten to one. Level eight would correspond to a top grade in any public examination a child may take at sixteen. Children in infant schools will usually be working at levels 1 to 3. Teachers will use the assessed attainments to help with planning and in making judgements about what sort of work a particular child would benefit from next.

WHAT ARE THE PROGRAMMES OF STUDY?

These set out the essential ground to be covered and the skills and processes which are to be taught. Each programme of study matches a certain range of levels which applies to all the attainment targets. The levels to which these programmes of study correspond depends on which key stage your child happens to be in.

WHAT ARE THE KEY STAGES?

If your child reached the age of between 5 and 6 on 1 August 1991 they will have started key stage 1 in the following Autumn term. There are four key stages altogether:

Key stage 1	ages 5 to 7
Key stage 2	ages 7 to 11
Key stage 3	ages 11 to 14
Key stage 4	ages 14 to 16

Although each key stage covers a range of levels, your child has the right to achieve success at whatever level is appropriate for them whatever their age.

In science, key stage 1 places equal emphasis on the knowledge and understanding of science and the 'doing' of science. Teachers will be working to help children with the skills they need to develop their ideas. How do you define a key stage in a child's life? At each stage it is thought to be good practice to develop ideas that start with the familiar and from there to develop ideas which are applicable in other situations. The home then gives children some of the best starting points for many important ideas in science. Milestones in understanding are marked by many strange things: one minute's worth of TV, an accident, a holiday – few are directly related to school unless, with teachers, we help to integrate them.

HOW WILL SCHOOLS CARRY OUT ASSESSMENT?

Science, as we have seen, is essentially a practical activity. In carrying out practical activity, children will inevitably need to collaborate with each other in making sense of their discoveries. They will need to discuss them with others. They will be putting forward their own ideas, testing them, and so on. All these processes are part of doing science. Such processes, as part of children's competence, must also be assessed in schools under what the national curriculum calls 'exploration of science'.

It is also important to note that the emphasis between exploration of science and 'knowledge and understanding' is equally shared for key stage 1 and is slightly biased towards knowledge at key stage 2. The ways in which children shall be taught remains the responsibility (and privilege) of schools. Your child's school will plan activities to help children refine these skills. At home, you can also play a major part in providing experiences which extend and give opportunities to practise these skills.

HOW WILL ASSESSMENT BE CARRIED OUT NATIONALLY?

In order to set up a framework to formally assess children's progress, the government have published guidelines. These are designed to make the assessment arrangements as manageable as possible for teachers and others and to make reports on children's achievement in them easy for parents to understand. Assessment involves measuring children's ability and reporting the results of those assessments at the ages of seven, eleven, fourteen and sixteen. These ages correspond with the end of each key stage. At key stage 1, it is expected that pupils will be assessed in the three core subjects with the addition of technology.

There has been some debate over what form assessment on a national level should take. Some want written tests at a specified time for seven- and eleven-year-olds. Others say we need tests which cover a far greater range of competence than a written test could measure. Whatever form testing eventually takes, there are certain things which, as a parent, you should try not to lose sight of.

- Does the form of assessment that your child's school adopts help your child?
- Do you feel that it helps you understand how your child is doing?
- Is it a fair or accurate appraisal of your child's strengths and weaknesses?
- How does the school use it to help plan teaching in the school?
- Are there things about your child which it does not or cannot measure? If so what other ways might be available to help in this area?
- Most importantly, what are children's views about these tests? Does your child him or herself see them as being either important or useful?

HOW WILL ALL THIS AFFECT YOUR CHILD'S SCHOOL TIMETABLE?

Don't be surprised if you find that in your child's school, the word 'science' may not even appear on the timetable. Indeed, some of the more usual 'reading', 'spelling' and 'long division' subjects might not appear either. Instead, you might find topics and themes such as 'The Romans', 'Ourselves' or 'Transport'. These form the basis of the work children do both in science and other subjects and they are planned to offer a mixed and varied diet of educational themes. Teachers at your child's school should be ready and willing to discuss with you the way in which the curriculum in your child's school is planned and should welcome your involvement.

Resource List

REFERENCES

Abbott C ed. *Be safe!: Some aspects of safety in school science and technology for key stages 1 and 2* Association for Science Education, 1990.

Aicken F *The nature of science* Heinemann, 1991.

Curtis A *A curriculum for the pre-school child: learning to learn* NFER-Nelson, 1986.

Department of Education & Science *Science in the national curriculum* HMSO, 1989; 1990.

Durrant J R 'Copernicus and Conan Doyle: why should we be concerned about the public understanding of science?' *Science Public Affairs* 1990; 5 (1): 7–22.

Harlen W & Jelly S *Developing Science in the primary classroom* Oliver & Boyd, 1989.

Hodgson B & Scanlon E *Approaching primary science* Harper & Row, 1985.

ILEA *Helping children to become scientific: primary science guidelines* Harcourt Brace Jovanovich, 1988.

Kerry T & Tolitt J *Teaching infants* Blackwell, 1987.

Marshall J & Flynn F *Science in the primary school: a guide for teachers* (BBC school broadcasts and the curriculum) BBC, 1989. op.

Neal P & Palmer J *Environmental education in the primary school* Blackwell, 1990.

Raper G & Stringer J *Encouraging primary science* Cassell, 1987.

Robinson M *The early years – a curriculum for young children: Science* Harcourt Brace Jovanovich, 1990.

Russell T & Watt D *Growth* and *Evaporation and condensation* (Primary SPACE project research reports) Liverpool UP, 1990.

Williams L P *Michael Faraday: a biography* Chapman & Hall, 1965; Da Capo, 1987. op.

Note: op. signifies that a particular book is out-of-print but may be obtained from a library.

READING AND VIEWING LIST

In a book of this size it is impossible to be comprehensive, so we have tried to be selective. Three main categories of books should prove helpful – information books which contain ideas and information about science; activity books which list activities and experiments to carry out using simple and easy to obtain equipment; and field guides or similar which help you explore the natural environment and identify the plants and animals living there.

GUIDE BOOKS

These guides to the natural world come in all shapes and sizes. If you feel that you and your child are at the same level of sorting and naming you would be best to start with the Usborne guides. The ideal book to have as a basic guide is the *Usborne nature trail omnibus* which includes birds, wild flowers, seashore, insects, ponds and streams, trees and leaves (Hart M, Usborne, 1978).
Cork B & Bramwell M *Nature trail book of rocks and fossils* Usborne, 1983.
A useful book if you or your child are developing an interest in geology.
Leutscher A *Animals, tracks and signs* Usborne, 1979.
Useful in helping you and your child develop your powers of observation even further.
BELS Biology Group *Biokeys* Blackie & Sons (Glasgow), 1988.
Not as colourful as the Usborne books but comprehensive and provides easy-to-use keys for identifying commonly found plants and animals. Excellent for keen beginners. Only available direct from the publisher.

More advanced guides are available. Hamlyn publish a good selection. The most outstanding and most comprehensive guides are the *Collins field guides*. The *Collins pocket guide to the seashore* is a must if you live close to or visit the seaside.

For minibeast enthusiasts there is a very good pair of minibeast posters which can be obtained from Harcourt Brace Jovanovich, High Street, Foots Cray, Sidcup, Kent DA14 5HP.

BOOKS FOR PARENTS

Sullivan M *Parents and schools* Scholastic, 1988.
Although this is a book for teachers, it could be handy for parents wanting an insight into schools' responsibility to parents and the ways teachers meet those responsibilities.
Holzinger PR *The house of science* John Wiley, 1990.

An introduction to all aspects of science. Questions are asked, answers are provided and activities suggested. Good if you want to get to grips with basic concepts in science and do something to explore them.
Lockhart G *The weather companion:* an album of meteorological history, science, legend and folklore John Wiley, 1988.
Provides a good background to watching the weather.

Bohren CF *Clouds in glass of beer: simple experiments in atmospheric physics* John Wiley, 1987.
A book in which quite advanced scientific concepts are explained.
Shaaf F *Seeing the sky:* one hundred projects, activities and explorations in astronomy John Wiley, 1990.
O'Connor M *How to help your child through school* Harrap, 1990. An excellent book which deals with general issues of your child at school.
Watkinson A *Primary Science – a shared experience* The Association for Science Education (ASE) College Lane, Hatfield, Herts AL10 9AA.
A resource pack to help teachers and parents make the most of science.
Harris C *A green school* (Bright ideas) Scholastic, 1991.
Although written for teachers there are lots of activities to do in here for young people concerned about the environment.
Auckland G & Coates B *Take nobody's word for it book of experiments* BBC, 1989
A handy book of recipes and experiments to do using easily available equipment. You may have already seen some experiments performed on television in the BBC series of the same name. Intriguing and easy to use.
Cash T *Science is child's play* Longman, 1989.
Extremely well written and worthwhile book to help you undertake science activities with children between the ages of 4 and 7.
Cornell J *Sharing nature with children* Exley, 1979.
An excellent book which provides inspirational suggestions for activities to help children empathise with nature. They are all well tried and tested and have proved very successful in stimulating children's interest.
Chinery M *The family naturalist* Macdonald and Jane's, 1977.
A classic activity book for those who are interested in natural history activities. In the same style there is also the *Family scientist* by Judith Hann (Macdonald & Jane's, 1979) and the *Family water naturalist* by Heather Angel (Peerage, 1986; Michael Joseph, 1982)

INFORMATION BOOKS FOR CHILDREN

The books we have selected here will help you to find out more about the topics covered in this book. If you cannot find these books in a library or bookshop you will certainly find others covering similar topics.

Cobb V *Lots of rot* A & C Black, 1988.
Extremely good little book which deals with all things to do with waste, rot and decay.
Newson L *Dealing with dirt: the science of cleaning* A & C Black, 1988.
An excellent information book which deals with very broad aspects of keeping

clean, health and hygiene, waste, rotting. Lesley Newson, the author, deals with difficult concepts and ideas in an imaginative and effective way. She has also written two other excellent books:
Feeling awful A & C Black, 1979 (about illness) and *Meatballs and molecules: the science behind food* A & C Black, 1984.
Bramwell M *How things work* Usborne, 1984.
Well written and colourful book on how everyday and not-so-everyday objects work.
Gans R *Rock collecting* A & C Black, 1984.
A book for younger children which looks at different rocks, how they are formed and what they are made of.
Nutkins T & Corwin M *Pets* (Factfinders) BBC, 1989.
Informative book about keeping pets.
Potter T *Weather* (Factfinders) BBC, 1989.
Everything you need to know about the weather including things to do.
Pollock S *Wildlife safari* (Factfinders) BBC, 1989.
How animals are adapted to their way of life.
Robbins R *Looking at nature* (Factfinders) BBC, 1989.
Information and investigations on the nature near you.
Parker S *Seashore* (Factfinders) BBC, 1990.
Exploring the seaside environment.
Pollock S *World in danger – water in danger* Belitha Press, 1990.
One of a series of six dealing with environmental issues. Includes water pollution in rivers and the sea.
Bronze L *et al. 'Blue Peter' Green book* BBC, 1990.
Very useful all round book about the environment and environmental issues.
Mayes S *Starting point science* Usborne, 1989.
A compilation book which looks at what makes it rain, what makes a flower grow, where electricity comes from and what is under the ground.

ACTIVITY BOOKS FOR CHILDREN 5–8 YEARS

Wilkes A *My first science book* Dorling Kindersley, 1990.
A large book which uses brightly coloured photographs showing all the equipment you need to carry out science activities and the activities themselves. The suggested activities are covered by many other books but the presentation is delightful and is bound to inspire you and your child into carrying out the activities.
Richards R *et al. An early start to science* Simon and Schuster, 1987.
A comprehensive range of activities packed into colourful pages.

Richards R *An early start to nature* Simon and Schuster, 1989.
In the same style as the above book but concentrating on nature and wildlife. Again well presented with lots of ideas and suggestions.
Bennett J & Smith R *Bright ideas for science* from Scholastic magazines Scholastic, 1984.
A compilation of ideas and activities based on science in primary schools. Lots of good ideas and explanations with good questions asked around the activities.
De Boo M *Bright ideas for early years* – science activities Scholastic, 1990.
Many original and interesting ideas for young children to do.
Fun with simple science Kingfisher Books
Well thought out and colourful books providing information, suggestions for things to do and examples which show the science in a real context.
Taylor B *Floating and sinking, Machines and movement, Shadows and reflections, Sound and music* Kingfisher, 1990.
Waters G *Science surprises* Usborne, 1985; *Science tricks and magic* Usborne, 1985.
Fun science activities demonstrated by a gang of colourful monsters.
Edom H *Science activities* Usborne, 1990, 1991.
Includes activities with water, with magnets and with light and mirrors.
Ardley N *My science book* a series of four published by Dorling Kindersley, 1991.
Subjects include air, colour, light and water. Attractive and informative.

ACTIVITY BOOKS FOR CHILDREN 9–11 YEARS

Richards R *101 Science tricks* Simon & Schuster, 1990.
A compilation of science activities based around looking, moving and paper.
Wilkes A *Simple science* Usborne, 1989.
Explains the basic principles of physics with simple experiments to try.
Claridge M *Living things* Usborne, 1985; pbk 1990.
Packed full of information and activities imaginatively and colourfully presented.
Bingham J *Usborne complete book of science experiments* Usborne, 1991.
A complete book of science experiments, lots of activities, well presented and well worked out.
Horn M & Orchard A *Search out science* book 2 Planet Earth and Moving about Longman, 1990.
Written to support the BBC school series *Search out science*. Colourful presentation and packed with good suggestions.
Parker S & Cash T *Weather* (Fun with science) Kingfisher Books, 1990.
Lots of activities, things to make, experiments and tricks, all provided with real life contexts. Colourful presentation.

Walpole B *Fun with science: Air, light, water and moving* Kingfisher Books, 1987; *More fun with science: Sound, electricity and magnets, simple chemistry and weather* Kingfisher Books, 1990.
Parker S *Weather* Kingfisher Books, 1990.
Hann J *How science works* Dorling Kindersley, 1991.
Fascinating projects attractively presented—an excellent book.
Ontario Science Centre *Food: the inside story* Cambridge University Press, 1986.
Extremely well presented and thought out book which provides information and lots of activities to do around the theme of food. Fascinating and fun.
Ontario Science Centre *The amazing science amusement arcade* Cambridge University Press, 1986.
Written by the same authors as the above book. Again extremely well thought out and clearly presented.
Cash T *Getting ahead in science* Longman, 1990.
Lots of questions to raise with older children, followed up with fascinating ways of exploring the answers through activities. Very well written in an accessible style.
Wade W & Hughes C *Inspirations for science* (Bright ideas) Scholastic, 1991.
A comprehensive selection of activities to undertake with children.

TELEVISION PROGRAMMES

Thinkabout Science
BBC 2.
For 5–7 year olds.

Watch
BBC 2.
For 6–7 year-olds

Science Challenge
BBC 2.
For 7–9 year olds.

Zig Zag
BBC 2.
For 8–10 year olds.

Search Out Science
BBC 2.
For 9–11 year olds.

All Year Round
ITV.
For 6–8 year olds.

Science Starts Here
ITV.
For 9–11 year olds.

Further details are published in *Radio Times, TV Times* and from *Ceefax.*

USEFUL EQUIPMENT

FROM SUPPLIERS

Science is a practical, 'doing' subject and for most activities (and certainly the activities in this book) you can easily make do with simple household materials. You may, however, find it useful to invest in some slightly more technical pieces of equipment too. A useful list would include the following: a pair of good magnets; litmus or pH papers; a 1.5 Volt battery (not rechargable); thin electrical wire; bulbs and bulb holders for 1.25 Volt light bulbs; a × 10 hand lens or magnifying glass; a pair of plastic forceps; a thermometer or liquid crystal thermometer strip; flexible mirrors.

Here is a list of equipment suppliers many of whom make their own products and sell those of other manufacturers, mainly to schools. We have indicated particular examples of equipment which may help you but it is advisable to ask suppliers directly for details or to send for their catalogues.

Note: Traditional chemistry sets can be quite expensive because you are paying for materials which may be difficult to obtain elsewhere. They are probably more suitable for older children of around 10–13 years who are more likely to make use of them. The actual science which comes out of them, although it may result in exciting smells and colours, is generally not much more sophisticated than the mixing and experimenting children can do with ordinary materials such as vinegar and salt.

Invicta Education. Invicta Plastic Limited, Oadby, Leics LE2 4LB. This company makes a useful transparent container called *Naturescope*, complete with lenses for collecting and studying plant and animal life. Also available is the *Young Naturalist* which is a large clear tray that can be used to scoop up specimens and examine them in transparent pots which have magnifying lids. Both these items are novel but quite expensive.

Osmiroid supplies a selection of specialist equipment including lenses and a *Weather Watching Kit*. Their catalogue can be obtained from Berol Limited, Oldmedow Road, Kings Lynn, Norfolk PE30 4JR.

Roopers Educational Suppliers, 20 Ridgewood Industrial Park, Uckfield, East Sussex TN22 5SX. This company supplies the Scientist ×20 Microscope, the *Pondwatch Kit* and *Acid Drop Kit* – the latter is of particular use for water testing activities. They also supply a range of magnifiers such as the *Nature Viewer, Box Magnifier* and various illuminated hand magnifiers and weather measuring equipment.

Palintest Limited, Palintest House, Kingsway, Team Valley, Gateshead, Tyne and Wear NE11 ONS. Can provide specialist water testing kits. The pH and nitrate pocket kits are the most appropriate for water pollution experiments.

C.E. Offord, Ticehurst Road, Etchingham, East Sussex TN19 7QT. Provides various viewers and is the main supplier for the robust Scientist ×20 Microscope (also known as the Offord Microscope). This microscope, although only twice as powerful as a good hand lens, has the advantage that children can put any object they find under it, explore the different parts, yet still recognise what they are looking at. It's a very worthwhile microscope for young children.
Heron Educational, Carrwood House, Carwood Road, Chesterfield S41 9QB. Provides various pieces of equipment including *Weather Watching Kit* and *Paper Making Kit*.
Galt Toys, Galt Educational, Brookfield Road, Cheadle, Cheshire SK8 2PN. Supplies a catalogue listing a considerable range of relevant equipment.
Pictorial Charts Educational Trust, 27 Kirchen Road, London W13 0UD. This organisation provides a range of colourful educational wall charts covering many aspects of science.
NES Arnold, Ludlow Hill Road, West Bridgeford, Nottingham NG2 6HD. One of the most comprehensive educational suppliers. The folding magnifier ×10 is obtainable from them.
Met. Check Ltd., PO Box 284, Bletchley, Milton Keynes MK17 0QD. Produces the BBC *Weather Watchers Kit*. It includes instruments, record sheets, maps and weather book.
National River Watch, Richmond Publishing, PO Box 963, Slough SL2 3RS. Produce a complete river pollution pack.

FROM TOY SHOPS

Most toy shops stock a selection of kits and equipment at a wide price range. When you do visit toy shops, try to make sure that you get a chance to see what's inside the kits before you buy.

Salter Science

Fun with Finger Prints contains materials that you need to take finger prints.
Fun with Crystals contains chemicals for crystal growing.
Fun with Magnets has a selection of magnets which can be used in different ways.
Fun with Colour. Things to do with colour separation.
Fun with Nature. Basic equipment needed to discover the amazing world of nature.
All of the above are appropriate for 8 years and over.

If you are prepared to spend more then for around £25 you can get Salter Science *Discover Microscopes* (suitable for 8-year-olds) which includes a microscope and prepared slides to look at. It is a very basic microscope and you will need additional slides. If you want a more advanced microscope, Salter Science also produce *Microscopes in Action* (Suitable for 10-year-olds) which costs more.

Salter Science also produce large chemistry sets such as *Chemistry in Action*, suitable for children aged 10 and over.

USEFUL ORGANISATIONS

Space prevents us from including a fully comprehensive list of organisations. Here are some of the more useful ones as a starting point. Remember to enclose a sae.

Association for Science Education (ASE), College Lane, Hatfield, Herts, AL10 9AA. This is the country's major organisation concerned with science education. The ASE produces several journals including the *Primary Science Review*. It also encourages the active involvement of parents in schools in school science through the special pack called *Primary science – a shared experience*, edited by Anne Watkinson. Contact them for further information.

WATCH Wildlife and Environment Club, The Green, Witham Park, Waterside South, Lincoln LN5 7JR. Encourages active scientific exploration of the environment. Children can join and become actively involved in worthwhile projects. They receive a regular colourful publication called *Watchword*.

British Association for the Advancement of Science, Fortress House, 23 Savile Row, London W1X 1AB. One of the leading scientific bodies in the country. The BA has a youth branch called the British Association of Young Scientists (BAYS). This is the major organisation for children interested in all aspects of science to join. It publishes a magazine for young members called *Scope*.

Department of Education & Science, Elizabeth House, York Road, London SE1 7PH. The Government department concerned with both education and science. Contact them if you want information about the various reports by H M Inspectors and the National Curriculum Council mentioned in this book.

Pre-School Playgroups Association, 61–63 King's Cross Road, London WC1X 9LL. The organisation concerned with playgroups. Their membership magazine *Contact* has activities which relate to science, and they also publish *Under Five* available from shops, again including material suitable for carrying out investigations relating to science with young children.

BBC Education, 201 Wood Lane, London W12 7TS. Produces schools and continuing education programmes as well as informative leaflets on current educational issues of interest to parents. Contact BBC Education Information for details on what series are coming up.

Independent Television Commission, 70 Brompton Road, London SW3 1EY. Also produce programmes to be used by schools.
Scholastic Publications Limited, Westfield Road, Southam, Leamington Spa, Warwickshire CV33 OJH. Publishes books and resources for primary education and also the magazines *Child Education* and *Junior Education*.

GOING ON A COURSE

Many university and college departments all over the country run short courses and day or evening sessions for members of the public. These are normally run through the college's Continuing Education Department or Department of Extra Mural Studies. The range of courses they provide is remarkable. There are a number of courses about science as the subject is becoming an area of increasing interest. Courses which present science in an interesting way are often very popular. One recent example is a course run by the University of Bristol's Department for Continuing Education on The Science of Toys. Contact Sue Pringle at Bristol University for details.

PLACES TO VISIT

There are now many places to go which both welcome children and provide learning opportunities in science. These include zoos, farm parks, city farms, botanic gardens and science centres. Also look out for museums that have science centres within them including some with live animals to study, or zoos which have small science centres.

- ask for activity sheets or at least a map—many places have an information point and produce worksheets for children to use;
- make the most of your child's interest in particular things; encourage them to make observations and give them new words to use;
- many places have special areas such as activity rooms which allow direct contact with objects—seek them out;
- some places have people called explainers—use them;
- visit the museum shop and take something home to do.

VISITING YOUR CHILD'S SCHOOL

Your child's school is a good point of contact if you need to find out more about science and ways you can help. Some schools have open days or open evenings to enable parents to do some of the science that their children do every day at school. These events are often very popular with parents.

ZOOS AND ANIMAL PARKS

There are a great number of zoos, wildlife parks, farm parks, butterfly farms and city farms around the country; all have considerable educational value. If you want more information about your local zoo the National Federation of Zoological Gardens of Great Britain and Ireland, Zoological Gardens, Regent's Park, London NW1 4RY can provide a free listing of member zoos with addresses, telephone numbers and some brief information about the zoos. Send them a stamped addressed envelope.

Some of the larger zoos which are particularly good for education:
Bristol Zoo, Clifton, Bristol BS8 3HA
Chester Zoo, Upton-by-Chester, Chester CH2 1LD
Edinburgh Zoo, Murrayfield, Edinburgh, EH12 6TS
Jersey Zoo, Les Augres Manor, Trinity, Jersey, Channel Islands
Marwell Zoo, Nr. Winchester, Hampshire SA21 1JH
Paignton Zoo, Paignton, Devon TQ4 7ED
London Zoo, Regent's Park, London NW1 4RY
Twycross Zoo, Atherstone, Warwickshire CV9 3PX
Whipsnade Park, Nr. Dunstable, Bedfordshire LU6 2LF
Wildfowl and Wetlands Trust (various centres around the country)
Headquarter at Slimbridge, Gloucester GL2 7BT.

For information about city farms send a s.a.e to: The National Federation of City Farms, Avon Environmental Centre, Junction Road, Brislington, Bristol BS4 3JP.

For information on farm parks send a s.a.e. to Rare Breeds Survival Trust, National Agriculture Centre, Kenilworth, Warwickshire CV8 2LG.

The national botanic garden is the Royal Botanic Garden, Kew, Richmond, Surrey TW9 3AB. There are others in Edinburgh, Oxford and Cambridge.

MUSEUMS

For information on museums, read the informative publication called the *Museums and Galleries in Great Britain and Ireland*, published by British Leisure Publications, Windsor Court, East Grinstead House, East Grinstead, West Sussex RH19 1XA and updated each year, which lists all the different museums in the country. This should also be available for reference purposes at your local library. If you need more specific information about museums, contact the Museums Association, 34 Bloomsbury Way, London, WC1A 2SF, telephone: 071–404 4767.

Some museums which are outstanding for their emphasis on science include: The Science Museum, Exhibition Road, London SW7 2DD. The national museum of science and industry (which has some unique material on show) as well as the interactive Science Centre called *Launchpad.*

The British Museum of Natural History, Cromwell Road, London SW7 5BD. Unique collections and excellent educational exhibitions plus a discovery room.

Other large museums in the UK with good scientific displays include: Birmingham Museum, Chamberlain Square, Birmingham B3 3DH; Liverpool Museum, William Brown Street, Liverpool L3 8EN; The Greater Manchester Museum of Science and Industry, Liverpool Road, Manchester M3 4JP; The Royal Museum of Scotland, Chambers Street, Edinburgh, EH1 1JF; The National Museum of Wales, Cathays Park, Cardiff CF1 3NP.

SCIENCE CENTRES

Essentially, a science centre is a place where you can explore and interact with exhibits and various pieces of equipment in a way that will help you and your child to learn more about scientific processes. Some science centres are based in museums, others are independent. Here is a list of centres:

Glasgow Dome of Discovery, South Rotunda, 100 Govan Road, Glasgow, G51 1JS
The Exploratory, Bristol Old Station, Temple Meads, Bristol BS1 6QU.
Green's Mill & Centre, Belvoir Hill, Sneinton, Nottingham NG2 4QB.
Hampshire Technology Centre, Romsey Road, Winchester, Hampshire SO22 5PJ.
Jodrell Bank Science Centre, Macclesfield, Cheshire SK11 9DL.
Launchpad, Science Museum, Exhibition Road, London SW7 2DD.
Light on Science, Birmingham Museum of Science and Industry, Newhall Street, Birmingham B3 1RX.
The Micrarium, The Crescent, Buxton, Derbyshire SK17 6BQ.
Science Factory, Museum of Science and Engineering, Blandford Square, Newcastle-upon-Tyne NE1 4JA.
Techniquest, 72 Bute Street, Pier Head, Cardiff CF1 6AA.
Xperiment, Greater Manchester Museum of Science and Industry, Castleford, Manchester M3 4JP.
The Centre for Alternative Technology, Llwyngwern Quarry, Machynlleth, Powys, Wales SY20 9AZ.
Armagh Planetarium, College Hill, Armagh, Co. Antrim, N. Ireland.

Index